D0849100

TOWARDS
HOLISTIC AGRICULTURE

Other Pergamon publications of related interest

TOWARDS HOLISTIC AGRICULTURE

A Scientific Approach

by

R. W. WIDDOWSON
Bourne, Lincolnshire

With a Foreword by
THE LADY EVE BALFOUR

PERGAMON PRESS

OXFORD · NEW YORK · BEIJING · FRANKFURT
SÃO PAULO · SYDNEY · TOKYO · TORONTO

U.K.	Pergamon Press, Headington Hill Hall, Oxford OX3 0BW, England
U.S.A.	Pergamon Press, Maxwell House, Fairview Park, Elmsford, New York 10523, U.S.A.
PEOPLE'S REPUBLIC OF CHINA	Pergamon Press, Room 4037, Qianmen Hotel, Beijing, People's Republic of China
FEDERAL REPUBLIC OF GERMANY	Pergamon Press, Hammerweg 6, D-6242 Kronberg, Federal Republic of Germany
BRAZIL	Pergamon Editora, Rua Eça de Queiros, 346, CEP 04011, Paraiso, São Paulo, Brazil
AUSTRALIA	Pergamon Press Australia, P.O. Box 544, Potts Point; N.S.W. 2011, Australia
JAPAN	Pergamon Press, 8th Floor, Matsuoka Central Building, 1-7-1 Nishishinjuku, Shinjuku-ku, Tokyo 160, Japan
CANADA	Pergamon Press Canada, Suite No. 271, 253 College Street, Toronto, Ontario, Canada M5T 1R5

Copyright © 1987 Pergamon Books Ltd.

All Rights Reserved. No part of this publication may be reproduced, stored in a retrieval system or transmitted in any form or by any means: electronic, electrostatic, magnetic tape, mechanical, photocopying, recording or otherwise, without permission in writing from the publishers.

First edition 1987

Library of Congress Cataloging in Publication Data

Widdowson, R. W.
Towards holistic agriculture.
Includes index.
1. Organic farming.   2. Agriculture.
3. Agricultural ecology.   I. Title.   II. Title:
Holistic agriculture.
S605.5.W52   1987   630   87-2314

British Library Cataloguing in Publication Data

Widdowson, R. W.
Towards holistic agriculture
a scientific approach.
1. Organic farming
I. Title
631.5'84   S605.5

ISBN 0-08-034211-6

Printed in Great Britain by A. Wheaton & Co. Ltd., Exeter

CONTENTS

Foreword

THIS book is exactly what its title suggests, a non-dogmatic textbook presenting, in recognized and accepted terms, the scientific basis underlying what is still usually, but misleadingly called "organic" farming. Nowadays there are many alternative adjectives in use, including the word "alternative" itself. Others are ecological, biological, and holistic.

In adopting this last, Mr Widdowson has, in my view, chosen the clearest and most accurate description of what the ecological alternative to intensive chemical farming really is.

Towards Holistic Agriculture is, as far as I know, the first textbook of its kind, and I welcome it most warmly.

In the current changing climate of academic, as well as practical, opinion it will fill a longstanding need. It should be in every school, college and university with an agricultural department.

It is clearly a must for teachers in alternative agriculture, for whom the demand already exceeds the supply, and I foresee that students in agriculture, horticulture, forestry and land-use generally, who have a scientific bent, will rapidly find it indispensable.

It will also fill another gap in existing technical books, for in his presentation the author has very skilfully linked the theoretical to the practical, so that the commercial farmer who is already practising ecological husbandry, or who wishes to do so, will find it an invaluable reference book, especially if he has the kind of enquiring mind that wants to know why this modern approach demands certain practices and bans others.

Those wedded to the dogmas of present-day orthodoxy in farming will probably query my application of the word "modern" to its alternative, but that is what it is. This fact has been very well expressed by the author in his short but admirable epilogue. "This new approach to farming", he writes, "is not . . . a return to the farming employed by our ancestors, it is a system which takes all the modern knowledge of the agricultural and other natural scientists, and develops practical methods of putting the knowledge into practice, ideally with no damage to the environment".

Not until I read this book did I realize how tremendously that knowledge has advanced since the days of my own practical and research experience. I found myself constantly astonished at the extent to which recent scientific discoveries have had the effect of interpreting and "justifying" holistic farming practices previously based only on traditional wisdom or trial and error – practices regarded, at best, as antediluvian, and at worst as bunk, by reductionists, as Mr Widdowson calls the orthodox zealots.

"Opposed to holism", he points out in his introduction, "is reductionism, a belief that complex phenomena can be explained in terms of something simple; this is the view generally held by chemical farmers." It is also the view, I might add, of many agricultural research scientists, an attitude which too often leads to the wrong questions being asked, which in turn leads to the wrong answers being received.

I have frequently noticed, when visiting agricultural research establishments at home and overseas, that when a district or individual farm suffers an outbreak of plant or animal disease, the investigating scientists usually tackle the problem by asking the question "why disease?" (meanwhile locating the symptoms!). Such researchers rarely study, or even visit, a nearby farm where no disease occurs, and there ask the question "why health?" If they did so they would be much more likely to arrive at the true causation of the neighbouring disease.

Not all research workers, of course, are guilty of this apparent lack of ecological understanding. I remember a case in point, when on one of the later P. A. Yeomans "keyline"-treated properties in New South Wales (referred to in Chapter 2) a serious trace mineral deficiency was initially diagnosed which Yeoman's technique showed to be due merely to an oxygen deficiency. This is not an uncommon phenomenon, the importance of which is uniquely (as far as I know) stressed in this book.

The late Sir Clunies Ross, the scientist noted for his pioneer trace element research in Australia, when he saw this demonstrated on the farm concerned, turned to the group of fellow-scientists and students who were with him, and said "Well boys, it looks as if we shall have to do our trace mineral work all over again."

Such open-minded willingness to admit that the wrong question may have been asked is increasing among scientists. This, plus the recent advances in scientific knowledge itself, is added to a growing awareness that there is no permanently sustainable future in maximum, as opposed to optimum, production which entails exploitation of land and livestock and ignores ecological reality, and this adds up to a new situation. It is *this*, I believe, that has led to the alternative approach, at long last, being taken so seriously.

True, we do not yet have a chair in Holistic Agriculture at any UK university, but a course has started in the subject at the University College of Wales, Aberystwyth. There is at least one such in Canada, and others in Holland, and the subject has a full university department with its own professor in West Germany at Kassel, so the general trend is such that the time is right for the publication of this book, and the book is right for the time.

I am quite confident that this is so in spite of the fact that, owing to belonging to the generation I do, much in the book is, to me, so new as to be beyond me. I was astonished to discover, among other things, the extent to which the latest computer technology is applicable, and potentially valuable, even to "traditional" farming. It is beyond my competence to comment on these sections of the book, my experience of such things being nil.

I do, however, have some experience in trying to think holistically, and this has taught me how different the solutions to practical problems can be when these are looked at and interpreted holistically. It has also taught me how terribly difficult it sometimes is not to lose sight of the wood for the trees. In the field of presentation, therefore, I can pay unstinted tribute to the way in which, throughout this book, Mr Widdowson has never allowed his reader to forget the unwritten 'w' in holism. He does not shirk detail; in fact the work that must have gone into researching and assembling the mass of data presented is prodigious; but however specifically and meticulously he may discuss scientific detail, and just as we begin to think that reductionism, i.e. simply solutions, are being offered, Dick Widdowson brings us back to a holistic approach with a sharp reminder that all details involved in biological functions are irrelevant except in the context of the whole. This is quite a remarkable achievement; another is that the book is so readable, a comparatively rare virtue in a textbook. In fact it is much more than a textbook and I wish it the wide circulation and success it deserves, which, in my view, it would be difficult to exaggerate.

Leiston, Essex
August 1986   EVE BALFOUR

Preface

IT was at Easter 1982 when I had been asked to speak at a school for farmers, held in the Kiewa Valley, Victoria, Australia, that I was first made aware of the demand that existed for knowledge on how to farm in a sustainable profitable manner. Farmers, smallholders and gardeners from all over Australia had gathered to hear speakers talk on tree planting, methods of tillage and different methods of animal and crop husbandry. In the seminars held each afternoon, under the red gum trees which line the river, I realized that here was gathered the most expert of stockmen and husbandmen who had travelled from every Australian state to glean theoretical knowledge to help them farm their enormously different farms, each with its own set of problems. There appeared to be no book to which they could refer which brought together modern scientific knowledge on allelopathy, the ethylene cycle, soil structure, and the use of industrial waste products, which suited their holistic philosophy.

In 1984 I returned to the Kiewa, and at that Easter School there were even more farmers attending. In a poll conducted by the organizers it was discovered that about 80% of those attending had come to try to find appropriate answers to problems on their units. Many were conventional farmers. The 1982 needs still appeared to exist, and they had not been met.

This book will, I hope, help all those who farm and who hold holistic ideals. I hope it contains some solutions to various problems – it does not contain any answers. The answers will be found and seen in the fields, stockyards and woodlands as each holistic farmer works out ways of applying the science in a manner appropriate to his unit.

There are of course many people to whom I, knowingly or unknowingly, am indebted. To all I give my thanks, but in particular I must remember my college lecturers, led by Robert Boutflour who taught me something of the science of agriculture and certainly taught me to develop individual thought; to the writing of Dr Schumacher where I was first introduced to holistic philosophy; to the Lady Eve Balfour who taught me how to apply my scientific knowledge to appropriate sustainable agriculture. There is the Wallace family who have encouraged me over the past 2 years to commit my ideas to paper: and finally, and not least, my thanks are due to my immediate family who acted as Devil's Advocate, who read the script to ensure that I was writing what I actually meant, and at the end typed the whole for my publisher.

Dyke, Lincs., 1986

Chapter 1

Introduction

HOLISM has been defined as the tendency in nature to form wholes that are more than the sum of the parts by creative evolution. Holistic agriculture is concerned in obtaining a correct grouping in farming systems which are in themselves sustainable; the organization to give this grouping within the farm and on a national basis and, it would be hoped, one day on an international basis, is of great importance.

The organization should be such as to give the fullest use of land resources, and the best utilization of available labour and capital. The aim is to provide a full diet of highly nutritious food at competitive prices for the consumer. At the same time there must be the realization that there is an interdependence of one farming system on another. This interdependence applies not only within the farm – livestock produce manure for root crops, root crops allow land cleaning to grow cereals for human food. The cereal crop needs fertility obtained from grassland which is eaten by the herbivorous livestock. This cycle can also be traced as existing between farms in differing parts of the country. The grass seed producer depends on farmers wanting grass seed to sow leys for livestock often obtained from the hill farmer, whose only real outlet to profit on poor land is in the sale of store cattle.

The labour usage in a well-balanced agriculture will be such that there is an even spread of work over the whole year, and so the farm worker will be an essential and vigorous part of a healthy and contented rural population, denied none of the material advantages of his urban cousins. The holistic agriculturalist should have no part in the "hire-and-fire" approach of the urban industrialist. The national economics of balanced agriculture should be considered on a holistic basis. The replacement of a farm worker by a machine must not be based on a simple cost-effective decision as to whether the depreciation on the machine amounts to less than the employee's total cost. The state must take into account the additional cost of unemployment benefit, the cost of housing benefit for the unemployed person's family and the loss to the exchequer of the workers's tax and National Insurance contribution.

Opposed to holism is reductionism, a belief that complex phenomena can be explained in terms of something simple; this is the view generally held by chemical farmers.

1

There is no argument between holistic agriculture and conventional agriculture about established scientific facts. For instance every agriculturalist understands the plant's need for nitrogen, but the method by which the plant is to obtain that nitrogen is fundamentally different. The holistic farmer will consider all the units, and do his utmost to ensure that none is damaged in his search for that nitrogen. The conventional reductionist farmer will apply the nitrogen in the most easily obtained form, usually ammonium nitrate, and will have little or no regard for the other factors which operate. Although he may nowadays care a little about soil structure he will give no thought to the fact that the very high levels of applied nitrate will give sappy plant growth which will then be more likely to be attacked by fungus so that he will have to use a fungicide. Nor will he care that the soil ethylene cycle is disturbed and he will need to apply phosphate and trace mineral fertilizers.

To understand the difference between the two different farming systems it is worth considering in detail two methods of crop production. The first, monoculture, is essentially primitive and reductionist, and consists of growing a crop of the same botanical family year after year. The system has the advantage that it requires the minimum of capital expenditure, and it tends to facilitate the sowing and harvesting of the crop, as all available labour on neighbouring farms can be centralized. However, the system leads to soil erosion by wind or rain because the successive crops lead to a deterioration in the humus content of the soil. In prehistoric times when the land had deteriorated it was abandoned; indeed some tribes still abandon land and only return to re-farm after a passage of time. In modern times the labour force becomes a shifting rather than static population (consider the combine harvester drivers of the U.S.A. who "cut" their way from the southern states into Canada) and this can be harmful to the family unit as well as the nation.

However, the greatest disadvantage is to be found in the great problems associated with weed and pest control. The balance of the ecosystem is upset, and the farmer has to resort to greater and greater inputs of fertilizers and persistent chemical pesticides, herbicides and fungicides to maintain worthwhile crops. Historically we have seen the fungi and insect pests becoming resistant to the particular biocide, and more money has continually to be spent on developing new effective sprays. All of this is taking place when the population at large is beginning to consider the necessity for such a high food input, and so we see the development of lakes or mountains of surplus food, all of which have to be supported by the taxpayer. The economic cost of monocultural crop production does not stop with the taxpayer, the cost to the farmer is clearly demonstrated in the University of Cambridge report[1] on farming in the eastern counties of England for the year 1979/80. The eastern counties are an area where continuous cereal growing is prevalent. In this report the author states, on page 32:

> The variable costs associated with growing a hectare of winter wheat in the harvest year of 1979 rose by approximately 24 per cent on the

previous year in current money terms, in real terms the rise was almost 10 per cent. The most noticeable rise was in spray costs which rose by over 12 per cent in real terms or about £10 a hectare in current money terms. Fertilizer costs also increased by almost 10 per cent in real terms or about £9.5 a hectare. At the same time yields declined by 3.4 per cent and the price per tonne declined by 5 per cent in real terms so that gross output per hectare declined by over eight per cent. It is not surprising that as a result of these changes the gross margin per hectare declined sharply by 13.3 per cent in real terms. The gross margin from a hectare of winter wheat is now in real terms roughly equivalent to what it was seven years ago. Although yields have increased substantially in the 1970's the variable costs of growing a hectare of wheat have gone up by 84 per cent in real terms. Thus looking to the 1980 harvest and beyond even if the real cost per hectare of variable inputs continues to rise at the same rate as it has over the last two years, then unless yields continue to increase at five per cent per annum the gross margin in real terms will fall. Indeed a twenty per cent rise in variable costs requires an increase of five per cent in yield to sustain constant gross margins, in the absence of any real increases in the price for wheat.

The 1983/84 report[2] from this same university department provides data to compile Table 1.1. The percentage changes are calculated as change in costs in real monetary terms (i.e. they allow for inflationary increases).

When the percentage yield figure exceeds the percentage gross output figure it must mean, allowing for stock differences which over a 10-year period could be ignored, that the value per tonne for that crop must in real terms have been reduced. It appears that in general the chemical farmer's reply to that condition is to increase his use of sprays and, for crops other than sugar beet, to increase fertilizer use. The one crop which is the exception is field beans, which for this 13-year period shows a reduction in spray costs. Field beans are not usually

TABLE 1.1 *Increase in yield, gross margin, fertilizer costs and spray costs from 1970/71 to 1983/84*

| Crop | Percentage increase | | | |
	Yield	Gross output	Fertilizer cost	Spray cost
Potatoes	+11.0	+83.0	+14.7	+60.5
Sugar beet	+15.3	−4.9	−10.4	+109.6
Winter wheat	+67.2	+35.9	+59.0	+349.2
Spring barley	+35.1	+21.1	+3.1	+87.0
Winter barley*	+35.1	+4.5	+53.2	+179.9
Field beans	+48.2	+45.3	−12.5	−14.9
Oilseed rape†	+5.6	+51.4	+117.6	+218.5

*Figure only available for the period 1973/4–1983/4.
†Figures only available for period 1972/3–1983/4.

PLATE 1.   *The problems of monoculture. Above*: Mount Fertiliser, Artesia, California. Over
30,000 m³ of cattle manure from intensive dairy unit. (Dr C. Tiejen). *Below*: Soil erosion
following heavy rainfall in Warwickshire (*Farmers Weekly*).

sown monoculturally and have been taken up by East Anglian farmers as what
is allusively called a "break crop".

It is dangerous to draw definite deductions from the information given in this
table, as the first and last year could distort the true trend, but the table does
cover a long period of time and appears to confirm the finding of the author of
the 1979/80 report: therefore some considerable confidence can be placed on
the conclusions.

The holistic cropping system depends on rotations. The idea of resting
agricultural land from continuous cropping appears to date from the times that
man actually settled land. The slash-and-burn methods of the wandering tribes

were no longer feasible, and in the book of Leviticus it is laid down as law that the land should be rested every seventh year. The cropping of the mediaeval village strip-farming system was wheat, barley or oats, fallow. This system allowed for the production of mainly food for human consumption, and as there was no production of food for livestock each winter saw the better breeding animals released to scavenge for what keep was available in the forest. Young stock was slaughtered and the meat salted down to provide for human needs in the winter months. The fallow year was essential in that it allowed for weed clearing, and under the fallow there was restoration of plant nutrient. It is interesting to note that even today if land becomes over-infested with weeds then fallowing is resorted to as the best method of cultivation, and the fallow is always followed by winter wheat, which in the U.K. is the best crop to utilize the increase in nutrient which occurred during the fallow period.

Closer consideration of the feudal cropping system shows that it in fact altered the ecosystem from grass to bare ground. In doing so it allowed weeds specific to the cereal crops to be reduced or eliminated. For instance the seeding of annuals and biennials can be stopped by the simple act of cutting off the flowering heads before seed formation – this practice is impossible in a growing crop.

The growing of the crop once in 3 years ensured that the species specific fungus and virus diseases could not be carried over; similarly insects, because they had no over-wintering host plant, were well controlled.

The introduction in the mid-seventeenth century of clover and turnips allowed for the development of the Norfolk four-course rotation – barley, clover, wheat, turnips. This cropping system was in fact the first scientifically designed rotation, and it held its position as the most important cropping system for over 200 years. Consideration of the system shows that a specific crop is grown once every fourth year, and that the same family of plants were never grown in successive years. Wheat followed clover, a fertility-raising crop, and the weed-fouling and nutrient-reducing crop, wheat, was followed by a root crop where good weed control was obtained. As it was possible to keep animals all winter, feeding them on roots and clover hay, farmyard manure became available in greater quantities than before for the root break. The keeping of livestock all the year round meant that fresh meat production assumed importance and planned animal breeding could be carried out; and the land was used more advantageously – half being for livestock and half for human needs. The system was almost ideal in so far as labour was concerned in that it gave a very even demand over the year.

The absence of mains drainage ensured that human urine and faeces were returned to the land, so enabling the whole farming and village system to be completely sustainable. Indeed the appearance of the village carpenter, cobbler and wool spinners and weavers made the village largely self-sufficient, and it is perhaps at this stage in our history that the political ideals of Plato were most nearly attained.

The Norfolk rotation was of course developed and modified not only according to soil type but also according to climate. In Essex, for instance, where the heavy clay soils would not allow winter stocking of cattle on the land, the root-cleaning crop was replaced by beans for human consumption – and the clover would be cut for hay either for sale to the London horse owners, or in some cases the crop was used for clover seed production. Another interesting rotation was developed on the Wiltshire Downs. It consisted of

wheat
barley
grass ⎫
clover ⎬ for sheep grazing
wheat
barley
beans
roots   for sheep folding

The rotation is in fact a double Norfolk four course, but it allowed 2 years to build up nutrients before growing the demanding wheat crop. The nutrient most in demand on the Wiltshire Downs was potassium, and by moving the sheep from one area of the farm to another, by "folding", the farmer was able to build up this very important nutrient without which it was impossible to grow worthwhile crops of cereals. It is little wonder that the phrase "the golden hoof of the sheep" came into the language, especially when the sheep which were kept were the "Down breeds" with their very fine wool.

The complement to cropping is of course the livestock unit. Holistic agriculture cannot exist without the use of livestock units. They are the processors of food unavailable to human beings into meat, milk or other products. They are the consumers of the growing quantity of human waste food products. By feeding on these waste products, livestock are returning to the farm some of the plant nutrients which have been sold off the farm, and their use helps in a small way to reduce the steady plant nutrient drain seen in present Western-style agriculture. Some of the livestock, such as poultry, feed on unsaleable grains which were produced on the farm, and give in return eggs and meat. Under proper husbandry methods they provide an aeration of grass swards by their scratching, and their devouring of insects whilst gleaning all helps to reduce pest problems. The livestock should not be kept as separate units designed to produce meat or eggs on small parcels of land, irrespective of welfare needs, where the products of their metabolic processes are a nuisance to be disposed of at the minimum cost. Indeed these waste products are an essential part of the whole which is appropriate agriculture.

Many readers would now ask whether holistic agriculture is not the same as organic agriculture? The answer to the question lies in the fact that holistic agriculture is a logical development, using modern scientific findings, of the agriculture described by Lady Eve Balfour in her book *The Living Soil* and which was, in spite of many adversities, confirmed by her work in the Haughley Experiment. Holistic agriculture accepts unreservedly the vital relationship between soil, plant, animal and man. However, as time has passed knowledge has advanced, and the subtleties of the interactions in ecosystems are now better understood. Companion planting, considered by many to be folk-lore when Lady Eve was writing, has become the science of allelopathy, and the elegant balance and vital importance of the ethylene oxygen cycle is now recognized. All these advances allow some of the older husbandry methods to be changed without in any way upsetting either the vital relationships or delicate balances. In light of these modern advances holistic agriculture would be more tolerant of the use of the chemical, urea, in adverse soil or weather conditions, as long as the amount used did not exceed about 1000 litres per hectare, and as long as the solution was not stronger than about 3% – the same level as is found in urine. It would, however, be much concerned at the use of poultry manure – wholly organic – which had been obtained from intensive poultry farms where the birds may have been fed with chemical additives in their feed which could well be lethal to soil micro-organisms; and if the manure had not been properly composted its use in the raw state would upset the ethylene cycle as surely as would the application of ammonium nitrate.

The newcomer to holistic agriculture may be forgiven if he does not readily understand why the application of chemicals by the conventional farmer is so abhorrent to the holistic farmer. After all, he will argue, the natural chemicals which are thought to be the essential part of allelopathy are so powerful that today, before they could be used, they would have to undergo endless testing before being given approval for use in farming (good examples are nicotine and atropine). The answer to this question is that the release is so very small in amount, and is localized in its area of activity. But more important these chemicals are not persistent, and can be absorbed into the environment of the ecosystem very rapidly; so the status quo is soon restored. Because the effect is so localized there is no real likelihood that resistance by fungus or insect pest can occur, and if a resistant pest were subjected to the chemical and survived, its genetic characteristic of resistance, being recessive, is soon lost in the normal population. Again why are the "droppings" of one hen on a piece of ground acceptable, whilst the application of tonnes per hectare of raw manure from hen batteries is not. The answer lies in the fact that the massive dose of ammonium cations applied when battery manure is spread cannot be dealt with by the clay domains, and plant nutrient balances are upset. If, however, the manure is mixed with other ingredients to give a good balanced compost the harmful concentration of nutrients is changed and reduced to a state so that the ecosystem can absorb them without any balance being upset.

However, perhaps the greatest divergence from present-day organic farming is the total rejection of a modern trend to lay down standards as to how animals or crops should be tended, irrespective of climate, soil type or topography.

Holistic agriculture is concerned with working ecosystems which can be as small as a field, and certainly not larger than the farm; it is concerned with balance in that unique unit and the imposition of national or international authority undermines the whole concept. None of us knows for sure what is right, we can only persevere to the best of our ability working "for" and "with": it is wholly negative and unproductive to be "against".

Because the subject of the book is holism, the book itself must be considered holistically. Chapter 2 deals with soils and their cultivation and Chapter 6 deals with grassland, but the one depends upon the other, as indeed do all chapters in this book. The order of reading is therefore in the hands of the reader. Holism is about forming wholes that are greater than the sum of the parts.

References

1. Murphy, M. C. (1981) *Report on Farming in the Eastern Counties of England 1979/80*. Cambridge University Press.
2. Murphy, M. C. (1985) *Report on Farming in the Eastern Counties of England 1983/84*. Cambridge University Press.

Chapter 2

Soil and Cultivations

SOIL lies at the boundary between air and rock or water – fresh or salt – and rock. It will therefore be an amalgam of each of the boundary materials. The importance attached to its composition will depend on which profession is going to use it. The engineer will be interested in its mechanical strength, the geologist in its possible use as a mineral resource. The farmer will have one of two views. The conventional chemical farmer will be interested in soil as a growing medium for his crops. These he will plant without regard to the effect on the soil, and feed and protect them with chemicals which will ensure that his plants grow and, by photosynthesis, use the energy from the sun to convert carbon dioxide and water vapour in the air into carbohydrates which are the most efficient source of energy for the plant (and also for the animals, including man, which feed upon the plants). This conventional farmer will, when adverse weather conditions prevail, show some consideration for soil structure as bad structures necessitate higher power requirements from his tractors (and therefore higher costs of production) to work them. On the other hand the appropriate farmer realizes that the soil is more than a growing medium; it is, as recently remarked by J. A. Wallwork,[1] "teeming with life. It is a world of darkness, of caverns, tunnels and crevices, inhabited by a bizarre assortment of living creatures". He will therefore tend his soil, paying attention to ensure that mechanical structure is maintained and that the living component is in a healthy condition so that the soil may remain the "living soil". He realizes that the soil through its airspaces should be a continuation of the atmosphere, by its water content a continuation of the river or the sea which it eventually borders, by its mineral fraction a continuation of the bedrock, and because it is a growing medium the plant residues will be incorporated by way of the plant and animal life residing in its structure to become a part of the whole called soil. It is indeed multidimensional and continually varying with time. Before considering the cultivation techniques which will be used to maintain this living mass it will be important to have a basic understanding of the function and make-up of the constituent parts.

9

2.1 Mineral Fraction

This is the skeleton of soil from which its strength is obtained. The fraction consists of particles varying in size from, say, large boulders to grains of sand and to particles so fine that they cannot themselves be seen even by microscope. In farming terms the husbandman is interested in all the particles. Stony soils can be difficult to work, especially if you are trying to grow potatoes, but as far as this book is concerned we shall only interest ourselves in those particles which are less than 2 mm in diameter. In the United Kingdom these particles are divided into three subclasses. Those between 2 mm and 0.02 mm are known as sand, those from 0.02 mm to 0.002 mm silt, and those smaller than 0.002 mm in diameter as clay. It is obvious from the foregoing that the varying quantities of the differing particles will give differing textures to the soil, and these differing compositions have given names to soil types: clay soils are those high in clay particle content: those with a high proportion of sand particles but with at least 30% of clay particles present are known as sandy loams. The mineral composition and texture of a soil will depend to a great extent on the rock from which it was developed or weathered. Sandstones give rise to soils with a high proportion of sand particles, whilst the granite feldspars tend to weather to give soils with a high clay content. Again, depending on the chemical composition of the original rocks will come some degree of the nutrient status – feldspars contain a considerable quantity of potassium and the clay derived from these rocks will be potash-rich when compared with silts which have been washed by water. In this respect it is worth noting that the alluvial soils found in river mouths or deltas are usually highly fertile as the river has, as it is slowed on approaching the sea, deposited nutrients and particles which in its early fast course it carried with ease.

Texture of a soil is of great importance; the greater the sand content the less water which can be held. Therefore these soils are more likely to dry out in summer conditions, and only plants and crops which can withstand these drier conditions will be grown by the appropriate farmer. Texture also has a dramatic effect on soil temperature and therefore the time of planting. Water has a specific heat which is about three times that of mineral particles in the soil, and the smaller the particle the more water it will hold. It follows then that the more clay particles there are in the soil the later is the soil in warming up in spring, because this type of soil will be almost waterlogged.

However, besides these structural and simple physical effects on the texture and handling ability of a soil, there is one other extremely important characteristic. This is known as the cation exchange capacity (C.E.C.). The surfaces of the clay particles carry a negative electrical charge and therefore attract any cations (positively charged ions) in their vicinity; at the same time they will repel any anions (negatively charged ions). These cations are freely exchangeable with cations which are in solution adjacent to the clay particles, and it is the measure of the soil's ability to exchange which is known as the C.E.C. The

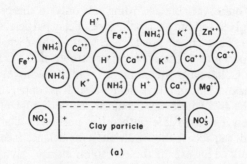

(a)

FIG. 2.1(a)   A good balance of plant nutrients is present in the soil. The clay particles are negatively charged at the face because the soil pH is about neutral.

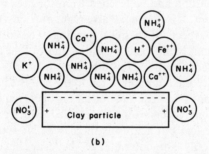

(b)

FIG. 2.1(b)   The application of ammonium nitrate fertilizer, or uncomposted animal slurries, increases the amount of ammonium ions in the soil and their presence in excess displaces other less frequently occurring cations.

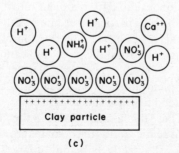

(c)

FIG. 2.1(c)   The ammonium cations are oxidized to the nitrate ions. The process increases the percentage of hydrogen ions, pH drops, and eventually the clay particles change their electrical charge. Cations will be leached, and anions, mainly nitrate, become attracted by the clay particle.

C.E.C. will be quite variable according not only to the percentage of clay particles in any given soil but also to the type of clay particle. Clays originating from vermiculites have a much greater C.E.C. than those which originate from kaolin. The importance to the appropriate farmer of the C.E.C. is that soils with a high C.E.C. have a greater ability to hold the positively charged cations of potassium, magnesium and trace elements such as copper, boron, cobalt and manganese in the area where plant roots are exploring for nutrients; as these cations are freely exchangeable with other positively charged ions in soil solution the plant can absorb these nutrients; the balance is correct and growth proceeds normally. However if the soil solutions are, say, almost saturated with ammonium ions (which are positively charged) – and this is the situation when soluble fertilizers based on ammonium nitrate are applied – then the most likely cations to become attached to the clay particle are ammonium ions, and so other ions at lower frequencies will lose their position on the clay particle, go into solution and be leached from the area where plant roots are exploring for nutrient. It is this removal of plant nutrients to lower levels which is one of the worst aspects of chemical farming. The correction to deficiency under these conditions being made by adding more of the deficient nutrient yet again upsets the balance required by both plant and micro-organism. Moreover by biological activity the ammonium ion in the soil is converted to nitrate (in which form the plant takes up most of its nitrogen requirement) and in the process of conversion hydrogen ions are released, which reduces pH. Thus not only is the balance lost but acidity is increased. Now the edges of the clay particles are generally amphoteric, i.e. they can be either negatively or positively charged. Whether they adopt a positive or negative charge depends on soil pH; they are positively charged at low pH – i.e. acid conditions – and negatively charged in high pH – alkaline conditions. Therefore to hold the beneficial cations husbandry methods have to be adopted to keep the soil slightly alkaline or, even better, neutral. The application of ammonium cations, as we have seen, causes hydrogen ions to be liberated and so makes soils acid; this is also undesirable from a plant nutrition point of view as the amphoteric edges of the clay particles become positively charged and can no longer hold the beneficial cation. The production of excessive quantities of nitrate, whether it originates from the large application of ammonium ions in chemical fertilizers or by natural processes, will upset the ethylene cycle which is an essential part of plant nutrition where holistic farming is practised. In a soil the clay particles tend to lie parallel to each other, and in this form they are known as domains. These domains are stable in the presence of calcium ions, i.e. when the soil is not acidic, and as the domains are larger than the individual clay particles the soil will effectively be full of larger particles and therefore more easily drained, so allowing access of air for better root respiration. Liming is the method used to keep the soil pH correct and at the same time keep calcium ions in the soil to strengthen the domains. The use of the relatively insoluble carbonate or sulphate salts of calcium which are used in

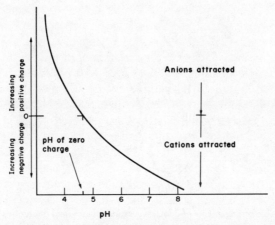

FIG. 2.2 Graph showing the amphoteric nature of clay particles. As the soil pH drops the
electrical charge gradually changes from negative to positive. The amount of charge and pH
of zero charge will vary with differing types of clay.

liming ensures that there is only a slow release of the neutralizing and
structure-aiding calcium cations. This slow release means that there is no upset
in the delicate plant nutrient balance.

2.2 Organic Matter

As with the mineral fraction, the organic matter in soils is a continuum from
freshly fallen leaves, stems and roots of gathered crops, known as the macro-
organic matter, through the partially humified material to the totally humified
material which is in reality a blend of complex organic acids. The humified
material is, like the clay particles, colloidal in nature. It is perhaps easier to
think of it as a sponge so that it has, for its particle size, a huge surface area, and
additionally it has a very high C.E.C. The C.E.C. of organic matter is
commonly twice the level of that of clay particles, and it therefore follows that
in those soils such as sands which are low in clay particles the decomposition of
organic matter to a colloidal state can increase the C.E.C. of the soil, and so that
particular soil's ability to hold exchangeable plant nutrient cations.

The colloidal organic matter has an ability to envelop the clay domains, so
making the particle size even larger. As has been discussed, the clay particles
will aggregate on their own in soils which have a high concentration of calcium
ions and are nearly neutral (pH 7) or at least not more acid than pH 5.5. Under
these pH conditions even greater aggregation and high C.E.C. will be obtained
in the presence of colloidal organic matter, because the C.E.C. of organic
material is improved in less acid conditions.

The organic matter in a soil is continually being changed, and it has been
estimated that, in theory, in British soils the turnover time for organic material

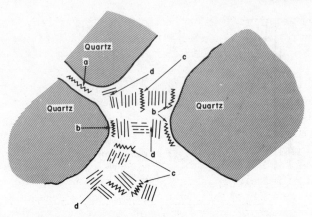

FIG. 2.3   Possible arrangements of domains, organic matter, and quartz in a soil crumb.
Type of bond: (a) quartz–organic matter–quartz; (b) quartz–organic matter–domain; (c)
domain–organic matter–domain; (d) domain–domain, edge–face. After Emerson[2].)

is about 18 years. This would indicate the importance of adding organic
material annually. More recent studies utilizing the radioactivity of carbon
indicate that the age of the carbon in soils is over 2000 years. It is probable that
this very aged carbon is more locked up than that which is more readily turned
over. It does, however, mean that it is essential that organic matter is applied
to the soil, so ensuring that the long-term carbon compounds are maintained.

So far discussion has only been about the use of the decomposed organic
compounds in adding C.E.C. and helping in the texture of soils. Some
attention must be given to the organisms which convert the living matter to the
colloidal state, as husbandry methods must take note of their presence and try
to provide conditions for their maximum performance. The organisms can be
grouped into three divisions: the vertebrate animals, such as rabbits and moles;
the invertebrates, such as slugs, earthworms and flatworms; and the micro-
organisms – yeasts, algae, protozoa, fungi and bacteria.

Vertebrates

The vertebrates are usually classed as pests by the farmer. The rabbits eat
his crops, and moles can severely damage newly germinated plants, although
their burrows can be of use in removing surplus water and their faeces do add
partially broken down organic material.

Invertebrates

Amongst the invertebrates by far the most important are the earthworms.
Their sheer weight in any hectare can be phenomenal – over 1500 kg has been
recorded in good pastureland, whilst in chemically treated arable land the total

weight may be as little as 20 kg per hectare. They feed exclusively on dead organic matter (not, as is often attributed to them, on the roots of living plants). They spend their time migrating between the soil and the surface litter layer, consuming vast quantities of the dead organic matter and mineral fractions of the soil. The two constituents are well mixed during their passage through the worm's alimentary canal, and then deposited as faeces in the burrows or as casts. It has been estimated that in a hectare of soil beneficial to worms the worms may consume a total of between 90 and 100 tonnes of clay and silt particles over a period of 1 year. As far as organic matter is concerned it has been estimated that up to 30 tonnes of cow manure could be consumed per annum by worms present in a hectare of land. The number of earthworms is of course mainly controlled by the quantity of dead organic matter available to them, but soil temperature, soil water and pH level all have an effect; in fact worms are rarely seen where soil pH drops below 4.5. The earthworm, with its ability to consume organic matter, rapidly incorporates the surface litter into the upper soil layers. Where there are no earthworms a clear boundary between dead surface litter and the mineral fraction can be noticed in the soil. Earthworms, besides being a major method of incorporating dead organic matter and of mixing organic matter with the mineral fraction, also provide through their burrows a major means for the access of air and the egress of surplus water.

Other members of the invertebrates – beetles, ants, termites, etc. all have varying effects on soil organic matter and generally feed on decaying organic matter found above the soil, having a beneficial effect, but in no way do they match the earthworm in importance.

In conclusion the molluscs, snails and slugs, must be mentioned; feeding on living material they are in reality pests and although some species live on fungi and vertebrate animal faeces, their importance as mixers of mineral fraction with organic matter is insignificant.

Micro-organisms

Under this heading are included the protozoa, yeasts, algae, fungi and bacteria. All play a part; indeed some are essential in soil life, and some understanding of them is necessary.

Protozoa – the smallest of the micro-organisms which feed on bacteria – are important in that they control bacteria numbers and in balanced conditions prevent any particular strain of bacteria gaining an upper hand.

Yeasts – some of the soil yeasts are known to produce antibiotics, and are therefore probably advantageous to general soil health. They are very efficient decomposers of lignin, but as they only thrive in non-acidic soils lignin decomposition cannot take place efficiently in soils which do not reach this non-acidic ideal.

Algae – the algae are photosynthetic, and therefore generally live on the soil surface. The "blue-green" algae can fix atmospheric nitrogen into amino acids, and therefore make a real contribution to the soil nitrogen level. The "blue-green" algae only thrive in alkaline soils whilst the green algae which do not fix nitrogen prefer acid conditions, so the maintenance of a pH approaching neutral is again seen to be desirable.

Fungi – other than yeasts – are either normal inhabitants of a soil, where they are highly beneficial, or spores may be blown into the soil where they germinate and start to feed on living tissue and are generally plant pests. The mycorrhizas are particularly important as they have developed a symbiotic relationship with other plants. There are two types of mycorrhiza recognized:

(a) The ectotrophic mycorrhiza are most commonly associated with trees, especially conifers. The mycelium wraps round the outside of the root, forming a sheath. These mycorrhiza store phosphorus in their mycelia, and these stores are released to the host when phosphorus stores in the soil become reduced due to intense biological activity. This release continues long after the host root has ceased as an active source for phosphorus uptake.

(b) The endotrophic mycorrhiza are found in most soils in the spore form. The spores only germinate when the root of a suitable host is close. The mycorrhiza grow best in a slightly acidic (pH 6.5) soil. The mycorrhiza hypha then invades the root either by way of the root hairs or directly into the root cortex, becoming very convoluted within the plant's root cells. In some cells vesicles develop and these are temporary phosphate storage organs. Some hyphae divide many times and form root-like growths called arbuscles. The arbuscle is the place where the fungus gives up the phosphorus to the host, and in return receives from the green plant energy in the form of photosynthesized carbohydrate. The external hyphae also form vesicles and spore bodies, the latter being fruiting bodies for release of spores to infect further root hairs when conditions are favourable. The presence of vesicles and arbuscles gives rise to the common name, V–A mycorrhiza. Practically all flowering plants can have a V–A mycorrhiza association.

Since 1966[3] it has been known that, where soil phosphate levels are low, plants uninfected with mycorrhiza will grow poorly or even die, whilst those with V–A mycorrhiza infection will thrive. It has also been shown that growth differences between infected and non-infected plants were greatest and in favour of infected plants when the phosphate was in the *least* available form. This 1966 work was confirmed in 1967[4,5] when statistically significant growth differences were easily obtained when rock phosphate was used, but not when highly available sources such as superphosphate were used. In all cases even where no significant growth differences were detectable mycorrhizal plants contained a higher phosphate content than the non-mycorrhizal controls. A

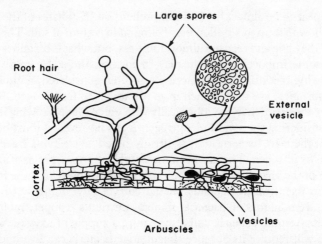

Fig. 2.4 Diagrammatic representation of vesicular–arbuscular mycorrhiza (not to scale). The fungus can be regarded as a two-phase system with mycelium in the cortex connected to external mycelium in the rhizosphere and soil. The first structures to be formed on internal mycelium are profusely branched arbuscles, and these are shown in the infection on the left. That on the right represents a later stage where vesicles are forming and arbuscles are being digested. Penetration may be via root hairs or epidermal cells. The basic part of the external mycelium is coarse non-septate hyphae on which finer rhizoidal hyphae are formed which are frequently tufted and septate. Vesicles are present on external hyphae. Thick-walled spores develop on the external hyphae. Two are shown, with that on the right containing oil globules. (After Nicolson[4].)

similar finding has also been noted for potassium, iron and copper in apples. Work carried out at this same time shows that mycorrhizal root infection reduces sharply as the level of available soil phosphate increases. In view of the foregoing it would appear unnecessary to waste energy turning insoluble rock phosphate into soluble forms; indeed the holistic farmer will not miss the point that in a soil with V–A mycorrhizal activity where composted material (and therefore phosphorus) is routinely returned to the land there is only a low requirement for the addition of rock phosphate. Of course the higher phosphate content of the plant is important when consideration of the phosphorus requirement of animals is taken into account. The presence of a V–A mycorrhiza on a crop plant will, because of the external hyphae, increase the length of absorbing root by as much as 50%. Workers at Ryall near Sydney, New South Wales, have recently shown[6] that differing V–A mycorrhiza have differing abilities to mobilize soil phosphorus reserves, and the search is now on to produce more productive strains which can then be released into soils.

Bacteria – in soil bacteria may be aerobic or non-aerobic, or indeed facultative, i.e. those which are usually aerobic but which can, under conditions where oxygen has become unavailable, use oxygen fixed in simple chemical compounds such as nitrates. The numbers and types of bacteria in any soil would appear to be almost beyond comprehension. Estimates of the numbers present show that up to 4 billion may be found in 1 gram of soil. The methods by which they live and feed are almost limitless, but what is certain is that they all have some important part to play in making the soil a more valuable plant-growing medium. The bacteria most often considered, and perhaps of most importance, are those which have the ability to fix nitrogen. Most well known are those which live symbiotically with green plants growing in the soil. Before considering the symbiotic nitrogen-fixers however, it must be remembered that there are bacteria and, as already stated, algae living freely in soils, which can fix nitrogen. The most common are the aerobic species of azobacters and some of the anaerobic *Clostridium* species. The ability of these free-living bacteria to fix nitrogen depends on the correct environment, which must contain trace amounts of bacterial mineral nutrients, copper, molybdenum and cobalt. Aerobic bacteria will also require a supply of oxygen, whilst the anaerobic will require its absence. Facultative bacteria can live under either condition.

Studies under field conditions show that the amount of nitrogen fixed by free-living organisms in land farmed appropriately can reach up to as much as 100 kg of nitrogen per hectare per year, whilst in intensively chemically farmed land only about 10 kg per hectare is fixed each year.

2.3   Symbiotic Fixation

Alders, bog myrtles and oleasters all have symbiotic associations, usually with yeasts. They of course do not have great importance in agriculture, but should be considered for use on acid wet hill sheep walks or in horticulture for hedging.

The most important symbiosis is, however, the relationship between members of the leguminosae and the rhizobium bacteria. The amounts of atmospheric nitrogen which can be fixed is, as the following table shows, phenomenal:

	kg/ha/year
Clovers, varying species	55–600
Lucerne, varying species	55–400
Peas, *Pisum* species	40–170

An understanding of the conditions which govern the relationships is important so that they can be exploited by cultivation methods to give maximum nitrogen production. The bacteria invade the legume root hair and build up a body in the root cortex. This then modifies to form a nodule, apparently on the surface of the root. Inside the nodule the bacteria divide several times and then surround themselves with a membrane. These sacs are known as bacteroids and there will be several such sacs in each cell within the nodule. The bacteria

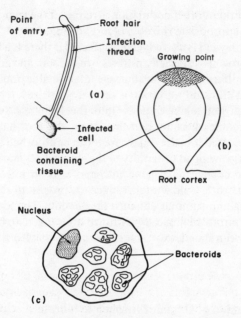

FIG. 2.5 **A**: Bacteria infect the legume through the root hair. An infection thread leads to a cell within the root cortex. **B**: Cross-section of an effective nodule. **C**: Infected cell showing nucleus and bacteroids. The bacteroids are enclosed by a membrane and nitrogen fixation can now take place. (After Nutman[7].)

within the bacteroid do not reproduce but they do require a plentiful supply of oxygen to respire. Nitrogen fixation takes place at the membrane surrounding the bacteroid, and a high proportion of oxygen at this site will reduce nitrogen fixation. The reduction of oxygen at the membrane, and the plentiful supply to the bacteria in the bacteroid, is maintained by use of a pigment in the membrane closely allied to haemoglobin. This pigment has an ability to take up oxygen at the membrane surface and can release it to the bacteria within the bacteroid. It is red and it imparts this colour to the nodule: nodules which are fixing nitrogen take on a pinkish hue. The energy needs for the fixation to take place are relatively high and the energy required by the rhizobia is obtained from the host legume. Besides a supply of air and energy, the fixation of the nitrogen depends on pH and a satisfactory supply of phosphorus, molybdenum and copper. In addition, invasion of the legume's root hair is poor if the pH is low or there is only a low level of calcium ions in the soil.

There are rhizobia which do not fix atmospheric nitrogen, and their presence can reduce infection by the useful rhizobia. There is evidence that there is a definite relationship between strains of rhizobia and species of legume. For instance, the rhizobia for lupins and lucernes do not appear to occur naturally in U.K. soils, and therefore if nodulation is required the legume seed must be inoculated by the appropriate rhizobia before sowing.

Nitrogen fixation yield is also adversely affected if there is a high proportion of mineralized nitrogen – especially nitrates – in the soil, and additionally if the host legume is suffering from a shortage of available moisture or is not photosynthesizing efficiently, fixation is severely impaired.

The nitrogen is not taken directly into the legume's system from the bacteroid. The nodules become detached from the root hair and the fixed nitrogen is then released into the soil where it can be absorbed by the host plant or by other plants growing in proximity to the sheared-off nodules. The rate of nitrification – the change of organic nitrogen salts to nitrates – is greatly affected by temperature. Cold winters reduce the process to almost zero whilst the warmer spring and summer enhance production. The ideal occurs when the production of nitrate biologically is immediately taken up by the plant and, as has previously been stated, excessive nitrate, even that from natural sources, is undesirable.

2.4   The Conversion of Organic Nitrogen to Nitrate

The complex nitrogenous compounds in organic matter are first of all broken down by enzymes, produced by soil fungi or soil bacteria, into simple nitrogenous compounds and carbon dioxide, and in the process energy for the use of the organism is released.

The simple nitrogenous compounds are then further broken down into ammonia with further energy release. This second stage is known as ammonification. The free ammonia in the soil reacts with carbonic acid or other organic acids to form ammonium (NH_4^+) salts. In temperate climates and under normal soil conditions this conversion to ammonium compounds proceeds continuously and ammonia is not found free in the soil. However, if large quantities of ammonium nitrate chemical fertilizers, raw slurry or uncomposted poultry manure, which are in themselves alkaline, are applied to the land then the pH of the soil will rise, the ammonia formed will persist as free ammonia or ammonium hydroxide and the pH will rise to above 7.0. When this stage is reached ammonium toxicity can occur and certain trace elements, especially boron, become deficient. Additionally mycorrhiza will not thrive and may indeed be killed out. The same situation can occur if soils are already neutral or alkaline, or excessive amounts of urea are applied to the land when, in addition to the normal production of ammonium carbonate, free ammonia is formed.

As the ammonium salts are formed further micro-organisms oxidize them to the nitrate state, in which form the plant takes up most of its nitrogen requirement. This last process, called nitrification, can only take place in well-aerated soils. This conversion to the nitrate state is an acidifying process

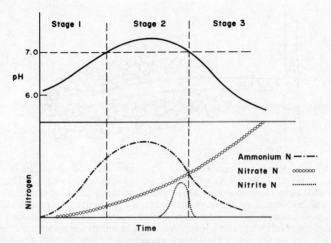

FIG. 2.6 Graph showing the changes which occur when heavy applications of raw animal slurries or ammonium nitrate are applied to nearly neutral or low buffered soils. *Stage 1*: ammonium nitrogen released or produced; pH rises. *Stage 2*: more ammonium nitrogen produced, the pH rises above neutral. Nitrates start to form and nitrites may be present. *Stage 3*: nitrate levels increase, ammonium level drops, and pH drops to a level below the original. (After Bunt[8].)

and soil pH will drop, very often to lower levels than that pertaining before the ammonium-rich source was applied.

Nitrites (which when present in water and food are associated with certain cancers) may also be produced when there is excessive ammonia production. Under conditions pertaining when over-heavy application of ammonium ions (i.e. uncomposted slurries or chemical ammonium fertilizers) are made nitrification can be inhibited.

If nitrification produces more nitrate than the plant can use then not only is the nutrient wasted (by leaching) but the ethylene cycle will be disturbed.

The holistic farmer will try to manage his soils and fertilizer application so that ammonification and nitrification proceed in balance, and then no upset of any of the soil biological processes will occur.

2.5 Ethylene Cycle

The limitation to plant growth is usually an inadequate supply of available nutrients. Many nutrients such as phosphorus are held immobile in the soil as complex ferric (oxidized) iron – Fe^{3+} – salts. These ferric salts have a large surface area, are highly charged, and tightly hold nutrients such as phosphorus and sulphate. In this form they cannot be leached from the soil but neither can they be absorbed by the plants.

When plants are established and growing well there is intense plant root activity, and in the area close to the root hairs there will be a proliferation of

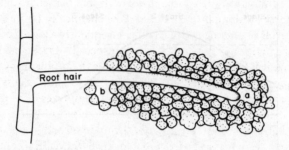

F IG. 2.7   (a) Microsite where intense root activity has used up all the immediately available oxygen. Anaerobic bacteria proliferate and ethylene is produced. The insoluble ferric salts reduce to ferrous salts. Plant nutrients previously "fixed" with the ferric salts are now available for take-up by the plant. (b) Intense root activity has ceased, and oxygen re-enters the microsite. The aerobic bacteria become active and ferrous salts are oxidized to the insoluble ferric state, and unused plant nutrient anions are again "fixed" and so no leaching occurs.

micro-organisms which will be feeding on plant exudates. This high activity will lead to a depletion of oxygen in the rhizosphere. Other micro-organisms, not dependent on oxygen, then begin activity and produce the gas ethylene which diffuses out through the immediate pore space. Once in these pores the ethylene inactivates but does not kill the aerobes. In a well-ventilated soil with the aerobes inactivated the oxygen level will rise and the ethylene level will drop, so allowing the aerobic organisms to gain the ascendancy. This cycle will be repeated continuously as long as the soil conditions are favourable. As the ethylene level in the soil increases the insoluble ferric salts are reduced to a ferrous – Fe^{2+} – state. The phosphorus and sulphur which were part of the complex ferric salt become available to the plant. Additionally the ferrous iron is attached to the organic clay domains, releasing cationic plant nutrients (ammonium, calcium, potassium, etc.) into the soil solution. As the anaerobic sites occur close to the plant root hairs where activity has been greatest then the nutrients are in exactly the right place for plant uptake. Once the aerobic activity restarts, the ferrous iron in solution is oxidized and the unused phosphorus and sulphur revert to the insoluble form and leaching is prevented.

Ethylene production, by its effect on the aerobic micro-organisms, regulates the rate of turnover of organic matter, and this further back in the chain aids the recycling of plant material and control of soil-borne plant diseases.

It is now known that ethylene production is severely restricted when nitrate nitrogen[9] is free in the soil, and this condition occurs in those fields to which ammonium nitrate fertilizers have been applied. Firstly the applied nitrate adversely affects ethylene production because it interferes in the production of the anaerobic areas, and secondly the ammonium nitrogen gives advantage to those bacteria which convert the ammonium salts to nitrate form, further increasing the free nitrates and creating an uneven balance in types of bacteria.

Ethylene gas is always found in undisturbed soils, but in modern agricultural soils the level is low or non-existent. These same chemically farmed soils also have declining levels of organic matter, with consequently a lower water-holding capacity and a deficiency of plant nutrients, especially trace elements. The conventional chemical farmer attempts to overcome the problems by applying fertilizers as a source of nutrients, and by irrigation to allow access of water which would normally be held in the organic clay complexes. The longer the practice of monoculture is followed the greater the input of water, nutrients and pesticides that is required. Restoration to undisturbed grass allows the ethylene–oxygen cycle to reassert itself and so bring in the natural controls seen in undisturbed land.

It is not clearly understood why ferrous iron is essential for ethylene production, but work at the Department of Biological and Chemical Research in Sydney[9] has shown that precursors which originate in senescent leaves on the surface of the soil – and which are amalgamated with the surface layers of the soil, probably by earthworms – react with ferrous iron and release ethylene. It is therefore easy to understand that, under undisturbed conditions of ley or woodland, precursors build up, whilst in land where stubbles are burnt off year after year precursors cannot accumulate.

The ethylene–oxygen cycle is therefore of great importance to the non-chemical farmer, but the conditions which lead to its occurrence are far more important in that they give rise to a sparing of nutrient loss.

2.6   Bulk Density of Soil

The bulk density of soil is the weight of oven-dried soil per unit volume. This, together with a known particle size, gives a method of determining the porosity of a soil.

The value obtained for any soil depends not only on the percentage of the varying soil constituents, but also on the degree of impaction of the soil. Values less than 1 gram per cubic centimetre would indicate very high organic content in the soil; those with values between 1 and 1.3 would be well-aggregated soils; and those rising to values of 2 would be for compacted clays or very sandy soils.

Bulk density is inversely proportional to the amount of pore space in a soil, and therefore it can be used as an indication of pore space. It is also inversely proportional to the organic content of the soil. So bulk density, which can be easily measured outside the laboratory, is a guide to the organic content of a soil, which is much more difficult to determine.

Cultivations must therefore be designed to ensure that the soil has good physical properties, a good nutrient supply and beneficial conditions for the proper functioning of the whole of the biomass.

The requirements consist of building good aggregates – the joining together of clay particles in the presence of calcium cations at a soil pH of nearly neutral (7) – by application of lime and organic matter. The aggregates will give good

TABLE 2.1

Soil property	Dependent on	Benefit
Good aggregate structure	Calcium ions; high organic matter	Rapid germination; no soil capping or soil erosion by air or water; improved drainage
Low bulk density	Calcium ions; high organic matter	Quicker spring soil warm-up; more soil air; good drainage of excess water but good moisture-holding capacity; vigorous root activity
Low concentration of ammonium and nitrate ions	No use of ammonium nitrate, Chilean nitrate or uncomposted chicken manure	Increase in earthworm activity; correct functioning of ethylene cycle; less build-up of hydrogen ions (less acidity)

aeration for root and micro-organism respiration which allows full exploitation of the soil by roots for nutrients and water. The organic matter gives more heat absorption, allowing quicker and more even germination. In addition, because the organic matter is like a sponge, it drains off excess water quickly but retains valuable moisture. Thus organic matter is vital for increasing the C.E.C. and water in sandy soils.

The application of a high concentration of ammonium or nitrate ions adversely affects the presence of trace mineral cations or proper functions of the ethylene cycle respectively. All the above desirable properties are found in both established woodland and permanent grassland, but as soon as cultivations are carried out there is a deterioration in soil physical properties. Indeed the decline is very rapid in the first year of cultivation, and then declines more slowly until it reaches a state where soil organic matter level is low, soil aggregation is poor and erosion by water or wind is common.

The cultivations of monoculture bring about this decline because the soils are directly exposed to rain and air for longer than in rotations. The surface litter, burnt off before the soil is turned by the plough, no longer exists, so allowing more rapid run-off (as opposed to drainage) of rain water and erosion is an increasing likelihood. Organic matter in the "turned up" soil is oxidized more rapidly, earthworm numbers decline because the stubble burn has depleted their food and the weight of the heavy cultivating machinery breaks up the clay domains, causing impaction. Very often the line of the plough sole produces a hard impermeable layer known as plough pan. The worsening conditions are reversed when the land is allowed to rest by returning it to grass.

Modern cultivations seem to have been designed as a method of weed control, and no consideration has been given to their effects on soil structure. Ploughing not only turns the trash from previous crops, but also places available nutrient present in the upper region of the soil profile lower down,

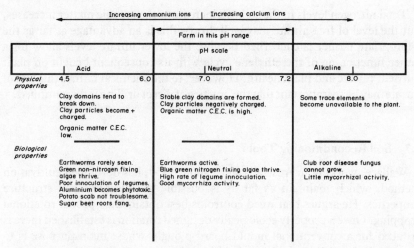

Fig. 2.8 Diagram showing the various physical and biological properties of soils at differing pH values.

often to depths beyond root exploration. Drawings of early ploughs show them to be no more than a pointed stick which, when pushed or dragged through the soil, just disturbed the upper few inches of ground. Modern ploughing also exposes the soil to the elements, allowing water erosion where plough pans have developed and also wind erosion in times of high wind speed. In sunny climates the exposure of the soil to the hot sun ensures that exposed organic matter is rapidly oxidized and lost from the soil. The loss of organic matter leads to poorer soil structure and in sunny countries where organic levels tend already to be low this loss can be disastrous. Subsequent discing and harrowing to prepare seed beds creates a fine tilth, but the passing of heavy machines breaks down the clay domains.

Tramline spraying methods used by the chemical farmer, though reducing the area of land which is compacted, does effectively ruin structure where the tramlines have been developed. Root crops are cleaned by inter-row cultivation but again at the expense of structure. Where tramlines run up and down even slight slopes, water erosion often causes deep gullies to form.

New ideas on minimum tillage or direct drilling are to be welcomed by the appropriate agriculturalist. With direct drilling the organic content of the top layers of the soil gradually increases and earthworms re-establish rapidly. The high biological activity, if not impeded by chemical inputs, and the increase in organic matter content so improve soil structure that potential erosion problems due to wind and water disappear.

On some soils after direct drilling has been practised there may be a drop in bulk density, but this is due to the disappearance of the large pores which are fairly rapidly restored by earthworm activity.

Total nitrogen levels in the soil will increase as the organic matter increases, but the level of free nitrate may well drop – this is an advantage as far as the appropriate farmer is concerned because the lower nitrate levels allow for a better functioning of the ethylene cycle with its consequent benefit on plant nutrient release and plant health. However, to the chemical farmer this loss of nitrate may be so important that the practice of direct drilling is discontinued.

2.7   Soil Reconditioning Tools

Wallace, working in Australia since about 1950, has developed cultivation methods which maintain as far as is possible the desirable soil structure properties. He argues that weed control is best obtained by proper rotational cropping. Once a properly ecologically designed rotation is established there is no need for a conventional mould-board plough, whose main purpose is to bury trash. In addition he has now shown that it is possible to have only one machine to prepare the land and to sow the crop at one time. His soil reconditioning unit is not a subsoiler, nor is it a chisel plough. The latter are designed to give maximum disturbance of the soil profile which will resettle fairly rapidly, the resettlement once again giving rise to soil compaction. The soil reconditioning tool is designed to give the minimum disturbance and to leave an air duct which allows removal of surface water and consequential ingress of air. The two factors give a more rapid soil warm-up in the spring with earlier sowing and potential increase in yields of cereal or grass.

The machine carries specially designed "boots" which pass through the soil at about 7.5–10 cm below the surface. As the boot passes through the soil it lifts the soil gently, and as it resettles leaves behind a duct triangular in cross-section. The duct gives access to a plentiful supply of air and the intense root activity which is always noted gives the correct conditions for the initiation of the ethylene cycle. Grass root lengths, where the tool has been used, are frequently over twice those of roots in untreated land. Experimental work in the U.K. and Australia has shown that where the tool has been used seed bed temperatures in the tunnel in winter and spring can be as much as 9 °C higher than soil temperatures at the same depth in untreated land, and it is this temperature difference, together with the slight condensation which occurs in the duct, which gives the excellent germinating conditions, as in a greenhouse. In summer the soil in the tunnel is cooler. These temperature differences are believed to be due to the presence of air, which acts as an insulator. In one experiment in the U.K. a grass and clover seed ley mixture had germinated and was showing above ground within 8 days of drilling, whilst the same mixture sown on untreated land took 3 days longer. In summer conditions, because of the insulating properties of the air in the duct, plant roots do not become excessively hot and dried out. The presence of such a plentiful supply of air

PLATE 2. *Above*: Wallace Soil Reconditioning Unit with seeder operating in grassland (I. Yule). *Below*: Plants removed from grassland. Plant on right with extensive root system from part of field treated with soil reconditioning unit. Plant on left from untreated part of same field (I. Yule).

PLATE 3.   *Direct drilling into grassland. Above*: Oats drilled into grassland (I. Fiedler).
Below: New grass seeded into permanent pasture (Hunters).

ensures sufficient nitrogen for rhizobial fixation at a high rate. This experimental work has further shown that when the machine is used in grassland, and where no artifical fertilizers were used, the dry matter content of the herbage had increased by about 11%,[10] and the actual yield of dry matter per hectare had increased by about 40%. This increase in dry matter content, as well as an increased dry matter yield of grass, confirms the observations made by stockmen that ruminants being grazed on treated land spend less time eating and more time resting than on traditionally managed grassland.

The percentage of phosphorus, magnesium and potassium in the herbage was also increased and consequently, as the dry matter yield improved, the levels of these same nutrients in the soil were reduced. This indicates that the intense root activity is allowing uptake of any available plant nutrient, reducing the potential for leaching losses. The greater amounts of magnesium and potassium can be explained by the fact that the very active root hairs are searching out the ions in the areas of the clay domains. The phosphorus increase is, in all probability, due to the efficient operation of the ethylene cycle, and also to the enhanced mycorrhizal activity which has been noted.

As we have seen, the presence of very friable soil in the base of the duct, the slight condensation which gives moisture in the duct and the higher soil temperature give ideal germinating conditions, and the attachment of a precision drill onto the soil reconditioning unit allows the possibility of direct drilling.

Since there is only one passage of tractor and implement over the area to be prepared, and that with the implement rear-mounted, the tractor never passes over prepared ground and the probability of breakdown of soil structure is reduced. Because the forward speed is slow and the tool only penetrates to a depth of about 10 cm, the tractor horsepower required is low, allowing lighter tractors. A seven-tine unit cultivates (and drills) at a rate of roughly $1\frac{1}{4}$ hectares an hour.

The latest development in the use of the unit has been to direct-drill cereals either for grazing or grain into grass swards. This has been enormously successful and the search is now to find suitable grass or clover species into which the cereal can be sown. In one set of experiments in Australia, in marginal cereal-growing country (less than 40 cm of annual seasonal rainfall), oats have been sown into a field of lucerne established in the previous year. Yields of over 3.5 tonnes per hectare were obtained. The important point in respect of this development is that the sward must not be excessively aggressive and should have a legume to provide nitrogenous nutrient. The legume must not grow tall enough to interfere with harvesting either by pulling down the cereal or, as in pre-war days in a wet year, making the grain very wet. Likely grasses appear to be meadow fescue, timothy or perhaps the rhizomatous Kentucky blue grass (*Poa pratensis*), and the legumes perhaps bird's-foot trefoil (*Lotus corniculatus*) or white clover. (The bird's-foot trefoil is of interest because it does not cause bloat in cattle.) The sward should ideally provide keep for cattle and sheep once the cereal crop has been removed, and this must therefore be another characteristic of the plants which are selected.

One of the greatest advantages of this system of cultivation is that the soil is never left bare, which means that neither wind nor water erosion will occur, and in hot or sunny climates the organic matter in the soil cannot be burnt up.

With this new approach to cereal growing the ecosystem has been moved from an arable one in which grasses are grown to a true grass ecosystem. The weeds associated with arable systems will disappear, and as the new field alternates between no grazing and grazing the weeds specific to one grassland type cannot become established. It could be argued that the continuous grass is a monoculture and this is a valid criticism, but if diseases and pests of the grassland become a problem then the land should be put into a root break. In those countries where the ultimate ecosystem is grassland no problem will occur since the ultimate ecosystem is, by definition, stable. The system is being developed for the growing of cereals (members of the grass family) in grass. It is equally possible to sow legumes (a species found in grassland) into worn-out pastures. Already in the U.K. cabbages are being grown successfully in a grass/legume sward.

The strongest ducts are created by the soil reconditioning unit when the land is slightly damp and where soil compaction or previous crop mat is heavy, i.e. when soil conditions for crop growth are, under traditional cultivations, at their worst. It is usual to use the unit twice in the first year of grassland renovation, and in subsequent years as often as root activity decreases. In well-restored pastures the use of the tool once every 3 or 4 years will be all that is needed. Where cereals are being grown the tool is, of course, only used once per cropping.

In fields where deep-rooted weeds such as docks are present special wide cutting boots can be used. The cutting of the weed below ground level soon has this vicious type of weed under control.

2.8 Keyline Cultivation

The formation of ducts when the soil reconditioning unit is used means that the tool can be used to move surplus soil water as required, either to drier parts

---

FIG. 2.9(a)   In the original Keyline system the whole area was surveyed to ascertain the contour at which the gradient changed. This contour was called the keyline.

FIG. 2.9(b)   Cultivations are carried out from the valley to the ridge. Only a slight fall is required.

FIG. 2.9(c)   (1) In dry lands a small reservoir R is contained by an earth dam D–D′, A track "T" is graded out. This serves as a farm road and also allows water to be led from the reservoir to the drier slopes. (2) This section across the track shows the fall given against the slope. Water can be flooded along the track and it will percolate through the soil and be distributed along the hill by way of the triangular ducts D. Note that by using this system the water is carried to the plants at root level.

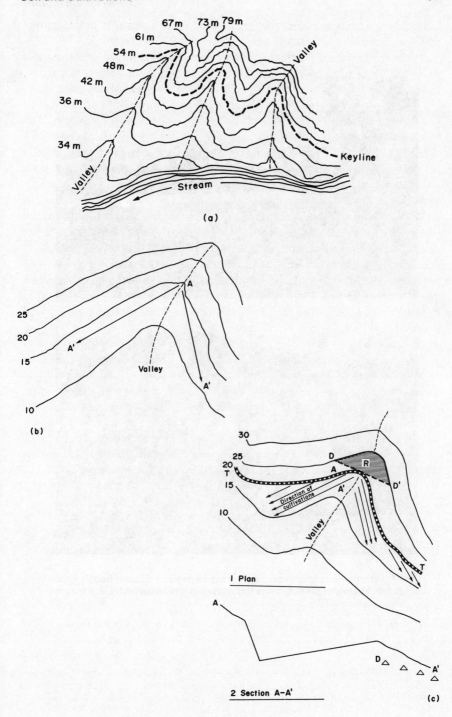

(a)

(b)

1 Plan

2 Section A–A'

(c)

PLATE 4. *Above*: Farm reservoir to supply water for keyline cultivated land (I. Fiedler).
Below: Volunteers planting native trees on keyline cultivated land (I. Fiedler).

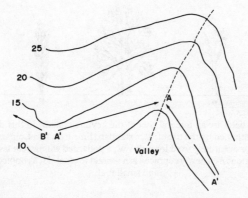

FIG. 2.10 Plan of reverse keyline developed by the author for wet British conditions. Water is conveyed from the ridge to valley, where natural drainage will take the surplus from the field.

of the field or to water courses. Wallace, working with the late P. A. Yeomans, used this drainage property to develop cultivation methods which have been called the Keyline method.

Under ordinary field conditions ridges in the fields are drier than valleys, and keylining is the method by which surplus water is moved to the dry ridges.

Originally Yeomans ascertained the contour on a field where the gradient changed, and this contour was known as the keyline. Cultivations with the Wallace tool were then made parallel to the keyline and surplus water drained from the valley to the ridge. The falls looked for were about 1 in 500. Under very dry conditions tracks graded into the hill were led from farm-scale reservoirs, made by damming the valleys, to the ridges. These reservoirs were contained by small dam walls made of earth, and were only about 3–4 metres high. The tracks acted as farm traffic roads, or as waterways which could be flooded from the reservoir, the water eventually finding its way into the previously Wallace-cultivated land and making its way to the dry ridges through the ducts. As the ducts are not waterproof there is leakage along the whole length and so water is gradually distributed at root level.

Nowadays less emphasis is placed on finding the exact keyline, but the field is cultivated from the valleys to the ridges ensuring that there is a continuous gentle downward slope. A spirit level is often fitted into the tractor cab so that the driver always realizes that he is maintaining the downward path. The tool should be lifted out of the ground at the ridge and, once past, reinserted into the soil to work upwards towards the next valley. Leaving this piece of unworked ground at the ridge ensures that there cannot be a large escape of water at this point, which could start water erosion.

In the U.K. where the problems can be very different the author has had success in developing a reverse Keyline cultivation method. Where land is

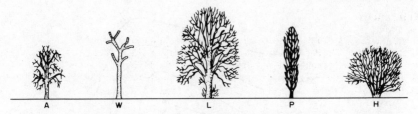

FIG. 2.11   Tree belts are an integral part of the keyline system. In this diagram the "A" section has been attained by the use of: L, a standard lime tree; P, a Lomardy poplar which can be coppiced or eventually used for pulp; W, a pollarded willow (cut last winter) grown for muka production; A, a newly coppiced ash – wood for fuel; H, a coppiced hazel for nuts and small fuel.

excessively wet – such as on hill land in the western parts of the U.K. – instead of having a fall from the valley to the ridge (which is the normal under ordinary climatic conditions) the fall is made from the ridge towards the valley.

Another part of the Keyline system developed by Wallace and Yeomans is the planting of tree belts. The species chosen for these belts were carefully selected native species which when mature give an "A" section. This shape of section allowed for air turbulence in the same way as hedges, with all the known advantages of pest control and wind speed reduction. The growing of tree belts allows the re-establishment of wildlife habitat with its importance to a true ecological balance, and is an eventual source of coppice or mature timber. It is suggested that the distance between these plantations should be at the same vertical height as the lower belt will reach at maturity, i.e. if the expected mature height of the lower first belt is 10 metres then the next belt would be planted at a vertical height of 10 metres above the first belt. Hence when the trees are at maturity the hillside, viewed from the frontal aspect, would appear to be wood-clad.

In flat lands these tree belts are planted to give wind breaks, so helping to stop wind erosion. They would in general be planted at right angles to the prevailing winds. Distance between the belts has to be a balance between aesthetics and the amount of land which can be spared for agro-forestry. Very often the trees will be planted where there are pockets of poorer soil. Indeed present-day uncleared woods already exist on this poorer land. When our ancestors cleared the woodland they only cleared where they knew the land was good. The advent of chemicals allowed more marginal land to be cleared.

The use of soil reconditioning units, the establishment of grass/legume mixtures into which can be sown cereal cash crops without the use of artificial fertilizers, the practice of Keyline cultivation methods and the growing of native trees in the manner described by Wallace and Yeomans is agriculture at its most sustainable.

References

1. Wallwork, J. A. (1975) The distribution and diversity of soil fauna.
2. Emerson, W. W. (1959) The structure of soil crumbs. *J. Soil Sci.*, **10**, 235–44.
3. Daft, M. J. and Nicolson, T. H. (1966) *New Phytol.*, **65**, 343.
4. Nicolson, T. H. (1967) Vesicular–arbuscular mycorrhiza – a universal symbiosis. *Sci. Prog. (Oxford)*, **55**, 561–81.
5. Murdoch, C. L. *et al.* (1967) Utilisation of phosphorus sources of different availability by mycorrhizal and non mycorrhizal maize. *Pl. Soil*, **28**(3), 329–34.
6. Smith, V. W. (1984) Personal communication.
7. Nutman, P. S. (1965) Symbiotic nitrogen fixation. In: *Soil Nitrogen* (eds Bartholomew and Clark), Ch. 10. American Society of Agronomy Madison U.S.A.
8. Bunt, A. C. (1985) Personal communication (Glasshouse Crop Research Institute, Littlehampton, Sussex). See also *Modern Potting Composts*, George Allen & Unwin (1976).
9. Smith, A. M. (1977) Microbial interactions in soil and healthy plant growth. *Australian Plants*, **9**(73).
10. Verity, A. J. (1985) *The Wallace Soil Reconditioning Unit*. Durham Agricultural College project.

Chapter 3

Sources of Plant Nutrients

UNDER any advanced farming system the production and sale of crops leads to a loss of plant nutrient from the farm, and unless steps are taken to return these nutrients to the farm a gradual decline in nutrient status will occur, with first a reduction in yield and then leading to trace element deficiency. Under conditions prevailing in a naturally occurring ecosystem the losses of plant nutrients in the system will be equal to the nutrient inputs. For instance, nutrients lost by the animals grazing are returned in part by the droppings and urine excreted by the animal, and in full when the animal dies and the body decomposes. Further nutrient return is obtained by the decay of plant tissues. The balance between the loss and inputs for all nutrients is equal. Figure 3.1 attempts to describe the balance in a simple manner.

Under existing conditions no attempt is made to recover from the consuming public the nutrients which are lost by way of household refuse or sewage. The deficiency due to these losses is made up by the application of soluble fertilizers. These are produced in the case of nitrogen usually from ammonium nitrate, which itself is a product of the very energy-intensive Haber process, and in other cases from minerals such as rock phosphate or kainitite. Not only does this practice place our chemical agriculture in a non-sustainable position, but it could be accused of taking their own natural resources from countries where minerals are mined. The present reliance on the use of nutrients imported to the farm either from the chemical or mining industries to make good sale losses is of such fundamental importance that the possibility of replacing these non-renewable inputs with other products must be considered.

The loss of nutrients from the farm by the sale of crops, livestock and livestock products can be calculated, and Table 3.1 estimates the quantity of major nutrients thus lost from U.K. farms in 1984.

It is of course only the nutrients sold in food destined for human consumption which are lost to the soils from which they are grown. An exception is sugar beet, which in all probability produces no nutrient loss to the farm. The leaves will be either fed to livestock or "ploughed in", and the sugar beet pulp, the by-product of the sugar-refining industry, is returned to farms as animal feed. The sugar sold to the consuming public is pure carbohydrate made by the plant from carbon dioxide and water taken from the air.

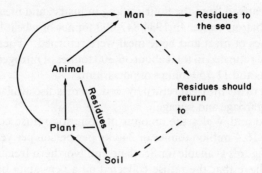

FIG. 3.1 Balance between loss and input of nutrients.

It must also be realized that the use of home-produced farmyard manure or liquid slurries in no way raises the nutrient status of the farm unless animal feeding-stuffs have been bought in; their use, however, does allow for the transference of nutrients from one part of the farm to another. This was well illustrated by the use of sheep on the chalk downs of Wiltshire, where sheep were allowed to graze on land too steep to plough, or on old pastures on the top of the hills during the daylight hours, but at nightfall they were moved to folds on the land to be used for next year's cereal production. Their excrement, especially urine, transferred potash, so essential on these naturally deficient soils, to the proposed new crop area.

Against the loss of plant nutrient due to sales off the farm must be added back the nutrients made available as animal feed by the return of products such as

TABLE 3.1 *Estimate of macro-nutrients lost from U.K. farms in 1984*[1]

Product	Kilograms of nutrient lost per tonne sold			'000s tonnes sold	Total nutrient lost (tonnes)		
	N	P	K		N	P	K
Wheat	18.0	3.8	4.5	3,523	63,410	13,390	15,850
Barley	17.5	3.7	4.9	1,872	32,760	6,930	9,170
Oats	16.5	3.5	4.2	117	1,930	410	490
Vegetables	20.0	3.7	5.0	4,870	97,400	18,010	24,350
Total crops					195,500	38,740	49,860
Beef	34.0	2.0	3.5	1,148	39,030	2,300	4,020
Mutton and lamb	23.0	1.73	2.5	286	6,578	495	715
Pig meats	24.0	1.92	3.4	916	21,984	1,760	3,110
Poultry	32.5	1.95	3.15	855	27,790	1,670	2,690
Milk	5.3	0.95	1.6	12,605	66,810	11,980	20,170
Eggs	19.0	2.2	1.4	605	11,500	1,330	850
Total livestock products					173,692	19,535	31,555

bran and wheatfeed from the flour-milling industry, and meat and bone meal made from abattoir waste. In 1984 847,000 tonnes of wheat by-products and 220,000 tonnes of meat and bone meal were returned.[2] These two classes of feed therefore returned in total about 36,300 tonnes of nitrogen, 17,800 tonnes of phosphorus and 11,750 tonnes of potassium.

In addition to the nutrient return by feed, there is also available the nutrients contained in garbage and sewage.

In England and Wales the amount of domestic refuse collected in 1984[3] amounted to 16.4 million tonnes or 340 kg per person per year. There are no comparable figures available for Scotland and Northern Ireland but there is no reason to believe that the refuse collected on a per capita basis differs from England and Wales, and using this assumption, the total U.K. domestic refuse would amount to 18.6 million tonnes. Domestic refuse is analysed for content on an annual basis and Table 3.2 gives the composition of U.K. garbage[4] for the year 1980 (the latest year for which figures are available). Therefore of the total 61.4% was of organic origin and so compostable, i.e. of the 340 kg of refuse collected about 209 kg is potentially available as plant nutrient.

Chemical analysis of this organic matter will vary, but it appears on average to contain 1.07% nitrogen, 0.51% phosphorus and 0.69% potassium[5]. It is therefore possible theoretically in the U.K. to recover 2.24 kg of nitrogen, 1.07 kg of phosphorus and 1.44 kg of potassium per person per year.

To these figures for domestic refuse must be added the waste collected from industrial and commercial premises. Statistics are not collected, but it has been suggested that the total in the U.K. may amount to another 36 million tonnes. Half of this sum coming from industry probably has to be ignored, as it is likely to be mainly metals or plastics and is likely to be contaminated with phytotoxic materials. The other 18 million tonnes from commercial premises consist mainly of waste paper, but the exception is hotel waste where as much as one-third of the total is in the form of waste food. As paper is mainly cellulose or lignin it is unlikely to have a high nitrogen, phosphorus or potassium content and should probably be composted with domestic refuse, making available humus for soil improvement but not adding much to the total nutrient recovery.

TABLE 3.2   *Composition of U.K.*
garbage (percentages)

Smaller than 20 mm screenings (ashes and cinders)	7.1
Vegetable and putrescible material	23.4
Paper and cardboard	33.9
Metal	7.1
Textiles	4.1
Glass	14.4
Plastic	4.2
Unclassifiable	5.8

TABLE 3.3   *Estimate of macro-nutrients recoverable from U.K. population in 1984*

	Recoverable per person per year (kg)			Total recoverable U.K. (55 million) population (tonnes)		
Source	N	P	K	N	P	K
Domestic refuse	2.24	1.07	1.44	123,200	58,850	79,200
Sewage	5.10	0.55	0.90	280,500	30,250	49,500
Total recoverable	7.34	2.97	2.64	403,700	89,100	128,700

Besides garbage the nutrients available from sewage must be considered. It has been calculated that the nutrients available per person per year from sewage amount to 5.1 kg of nitrogen, 0.55 kg of phosphorus and 0.9 kg of potassium.

Table 3.3 summarizes the situation in respect of recoverable nutrients in the U.K. Table 3.4 is a balance sheet for nutrients lost as against nutrients recoverable, and from this it is noted that there is a huge surplus on all nutrients. This surplus is probably attributable to the residues of imported food, paper and textiles. As has already been stated, the vegetable and putrescible content of domestic refuse is 23.4% of the total. In other words in 1984 over 4.35 million tonnes of domestic refuse was "swill". Only about half of this quantity is needed to restore the soil nutrient loss and the other half, which is equivalent to about 0.7 million tonnes of high-quality cereal, could be made available for animal feed. In the calculations no account is taken of the "swill" available from hotels. The question can be asked, why is this plant nutrient not recovered? The problem is that in the U.K. human excrement is mixed in the sewers indiscriminately with the waste products of industry, and

TABLE 3.4   *Balance sheet of nutrients lost and recoverable in the U.K.*

	Losses from farms (tonnes)		
	N	P	K
From Table 3.1			
Crops	195,500	38,740	49,860
Livestock	173,692	19,535	31,555
Total losses	369,192	58,275	81,415
Recoverable from the population From Table 3.3			
Domestic refuse	123,200	58,850	79,200
Sewage	280,500	30,250	49,500
Total recoverable	403,700	89,100	128,700
Surplus nutrients	34,508	30,825	47,285

the valuable plant nutrients become contaminated with metals such as mercury, cadmium, zinc, nickel, chromium, etc., all of which are phytotoxic. This therefore renders them useless in agriculture. The same is true of garbage, which is not separated from toxic or non-compostable materials before collection.

It is obvious that as sources of phosphorus and potassium minerals run out, due to over-mining, our attention will have to turn to the recovery of these waste nutrients from our own adequate resources. However, until this time occurs, and the appropriate agriculturalist would press for recovery sooner rather than later, he, like his chemical counterpart, will have to rely on mined minerals. He will, however, ensure that the amount of fertilizers which he purchases is minimal and that the fertilizers are those which need to undergo microbiological processes before release to the plant; also that his source of nitrogen will, where possible, come from the fixation of nutrients by legumes.

3.1 Fertilizers

Holistic agriculture is as concerned as any other method of farming to ensure that maximum yields are obtained under the conditions prevailing at any particular farm. The difference between them lies in their approach to nutrient supply. The appropriate farmer is always guided by answers to the fundamental question: "Is the application of any chemical input into the ecosystem going to be so great that the localized environment cannot absorb the input without a significant disturbance of that environment?"

In general terms this means that no fertilizer can be used which is so soluble that it increases the concentration of cations in the immediate vicinity of the clay organic complexes in the soil, or upsets the ethylene cycle. Thus the modern soluble fertilizers such as those based on ammonium nitrate are unacceptable.

Equally the highly soluble chicken manures, containing large quantities of ammonium salts, which are often freely obtainable from intensive hen units, would not be acceptable in their "as-received" form, although they are "organic" in origin. They are, however, admissible once they have been put through a composting system which would ameliorate the high solubility. On the other hand "chemicals" such as urea would be more acceptable (under strictly defined conditions – discussed later in the book) because before being capable of attachment onto the clay organic domains, or of absorption, they have to undergo biological change. Indeed if the localized environment cannot change the urea then the urea breaks down into ammonia which can rapidly escape from the soil environment.

All plants require as food materials not only carbon, oxygen and water as the basic raw materials of carbohydrate synthesis, but also many other elements, some at macro-level and others at the trace, or micro-level. Nitrogen, phosphorus and potassium are required in great quantities so that the crop may grow to give high yields from efficiently operating plants. The trace elements are essential to obtain proper growth – a deficiency gives rise to diseases such as grey speck (manganese deficiency) and heart rot (boron deficiency). They are usually found naturally in sufficient quantities to ensure that the appropriate farmer may not be concerned about their supply. It is only when the balance of nutrients is upset that deficiency is likely to occur, and this is most likely when the local soil environment is overburdened by application of large quantities of soluble fertilizers (as discussed in Chapter 2). A nutrient which has an important place in between the macro- and micro-nutrients is calcium. As we have already seen, it has a vital role in the formation of a good soil structure, but it is equally important as a plant nutrient.

3.2 Macro-nutrients

Nitrogen (*taken up as nitrate*)

Nitrogen is required by the plant to form the vegetable proteins. If applied to excess it will also increase the bulk and yield of the crop by the production of more leaves, although these leaves are less efficient as photosynthetic agents than those where excessive quantities of nitrogen have not been applied. The rapid growth which nitrogen gives the plant is often very "soft" and therefore more susceptible to plant and pest disease, the thinner cell walls being more easily penetrated by fungus or aphis, the carriers of virus infection. Again, excess of nitrogen will markedly delay ripening in cereal crops, and this can lead to severe problems at harvest time when autumn storms and heavy rain cause seed shed before the crop is ripe; additionally, the luxuriant growth is too much for a poorly developed root system, and will cause lodging.

Deficiency symptoms

Acute deficiency, rarely seen these days, causes weak spindly growth with pale yellowish-green leaves. It is usually seen where cereals have been monocultures, the organic content of the soil has reduced and a wet spell has leached out of the soil the soluble nitrates or ammonium salts.

Phosphorus

Phosphorus is required by the plant to enable it to establish a good root system at an early stage, and this is followed by good tillering in the cereals. Obviously its beneficial effect on root development makes it important in those crops such as mangolds, carrots, etc. where the final harvested part of the plant is its "root". It is also important for legume crops but this is, in all probability, due to the need of the rhizobia (the symbiotic nitrogen-fixing bacteria) for this nutrient.

Phosphorus Distribution

Phosphorus deficiency in the U.K. is associated with the oolitic soils of the Cotswolds and Yorkshire, the heavy loams of Suffolk, the peat soils of the Fens and the acid soils derived from the millstone grits.

Deficiency Symptoms

Plant growth tends to be spindly, and an examination of the roots shows a poorly developed system. Often the foliage of plants will be bluish-green. In grassland there will often be an absence of clover or other legumes.

Fixation of Phosphate and Mycorrhizal Activity

Only about 25% of phosphate applied in the soluble form is taken up by the plant. The remaining 75% is "fixed", usually as aluminium or iron salts, and will become available only after it has been released by the mycorrhizal activity. As fully discussed in Chapter 2, the mycorrhiza are a group of fungi which are common in all soils. They appear to grow best in slightly acid soil conditions (pH about 6.5). At one time it was believed that they were only associated symbiotically with the conifers, but as a result of recent work in Australia and New Zealand it is now known that they associate with practically all plants. They are capable of mobilizing mineral reserves of, particularly, phosphorus and sulphur from the soil, and these are given up to the host plant which in return provides the fungus with synthesized carbohydrate.

The farmer should ensure that this relationship, which is at its most active in springtime when the host plants have a high requirement of phosphorus, is not upset or stopped by overliming or by operating farm systems which will reduce the organic matter content of soils, reducing or even killing out the mycorrhiza.

Potassium

Potassium is required by the plant in the "manufacture and translocation" of the carbohydrates in the plant. It appears to increase the efficiency per unit area of the leaf, and is associated with the resistance of the plant to disease. This increasing efficiency is manifest under varying seasonal conditions. In dry seasons its presence will delay ripening and control the loss of water from the soil by reducing the crop's transpiration rate, whilst in wet seasons its ability to increase efficiency of the whole plant enables the plant partially to overcome the adverse weather conditions.

Potassium Distribution

Potassium is practically always present in the igneous rocks, and therefore soils which have developed from these rocks tend to be rich in potassium; such soils are the clays. On the other hand light sandy or calcareous soils will tend to be deficient.

Deficiency Symptoms

Cereals will show plenty of tillers, but they will bear few flower stems. In all plants there will be stunted growth, and leaf margins often show a scorched appearance. Because of the ability of potassium to strengthen the plant against disease, deficiency will often be associated with an increase in plant disease infestations.

Calcium

Calcium is required as a plant nutrient, but perhaps its most important role is when it is applied as calcium carbonate – "lime". Calcium carbonate has its place as a neutralizing agent of acid soils, so allowing not only better growth but also proper development of earthworms and microbiological organisms – all improvers of soil structure.

Lime is not, of course, required on the alkaline chalk or oolitic limestone soils as there is always a release of calcium cations from the weathering of the parent rock. However it will need to be applied regularly on those soils which overlie the igneous or sandstone rocks and which are not naturally alkaline. "Lime" is available as finely ground limestone (calcium carbonate). This material, being only slightly soluble, is the most commonly used source of calcium. Gypsum (calcium sulphate) is also a useful source when supplies of this mineral are freely available. Burnt lime (calcium oxide) and slaked lime (calcium hydroxide) were once favoured, as it was considered that their solubility was an advantage. To the appropriate farmer it is this aspect which makes them unacceptable. In certain areas far removed from sources of ground limestone reliance is placed on calcium-type sands which are found on the seashore. These sands can, however, have a high salt content which is injurious to soil texture. Again there is a ready demand, mainly by horticulturalists, for calcified seaweed. This product has a very high neutralizing content, and additionally can act as a supply of trace minerals. However at present the demand is outstripping the ability of the seas to provide sufficient seaweed, and the user should ask himself whether or not he is justified in exploiting the marine ecosystem for his own benefit.

Overliming can produce soils which are so alkaline that manganese, copper and zinc become unavailable and reduce mycorrhizal activity.

Magnesium

Magnesium is classed as a macro-nutrient although the requirement for differing crops is not as fully established as for the other macro-nutrients.

Except on naturally deficient soils magnesium deficiency, indicated by yellow strips running lengthwise on the leaves, is rarely seen. If a deficiency is seen the holistic farmer can overcome the problem using Dolomite limestone instead of ordinary limestone. Dolomite is an almost insoluble rock consisting of a complex salt of magnesium and calcium carbonates. Thus application of this material will give a slow release of magnesium cations and at the same time supply the soil with structure-benefiting calcium.

3.3   Micro-nutrients

The plant micro-nutrients (trace elements) are iron, manganese, zinc, copper, boron, molybdenum and chlorine. In appropriate agriculture it is rare to see deficiencies because the holistic farmer does not apply soluble chemical fertilizers which, as described in Chapter 2, upset the delicate balance in the clay domains, considerably diminishing these trace elements by leaching. In addition the return to the land of organic residues ensures that those trace elements which are surplus to the animals' requirements are returned to the soil. Some of these elements such as boron and zinc, although essential in trace quantities to prevent plant deficiency diseases, are phytotoxic if present in excess. So once again in practice the balance between the two extremes is vital. The trace elements in plants are very often the source of trace elements for animals. In New Zealand, where soils tend to be naturally deficient in molybdenum, it is noted that vegetarians who consume vegetables only grown on these soils develop serious dental caries, one cause of which is attributable to molybdenum deficiency. With these soil conditions it is very important to ensure that this trace element is built up in the soil. The holistic method is not to apply solutions of the deficient mineral salts which would upset the clay domains, but to use composts made from urban waste (uncontaminated with excess heavy metal). This urban waste will generally contain sufficient of the trace element to correct the deficiency. Another source is a fertilizer such as basic slag which contains trace amounts of this element.

Incidentally molybdenum deficiency can often be corrected by the simple aerating and draining of soils, as the oxidized forms of molybdenum are those which are most available to the plant.

3.4   Organic fertilizers

Farm Produced

Farmyard Manure

Storage of farmyard manure in heaps gives anaerobic fermentation, and it is probable that pathogens and weed seeds are not killed. About 12.5% of the organic matter is lost, but nutrient levels will be concentrated. Composting of the farmyard manure gives aerobic fermentation and pathogens and weed

TABLE 3.5   *Analysis of farmyard manure*

Dry matter	20–25%
Organic matter C : N ratio	20 : 1
N	0.2–0.6%
P_2O_5	0.1–0.7%
K_2O	0.1–1.0%

Note: only 30% of nitrogen becomes available in the year of application; 15% is available in the second year, and in each succeeding year the amount available is halved. Phosphorus and potassium are not "fixed" and are easily available to the crops.

seeds are killed. Farmyard manure builds up the organic content of the soil, with all its consequent benefits.

Urine

Urine is a quick-acting manure which is a nitrogen–potassium fertilizer. It does not build up the humus content of the soil. The application of urine in warm weather can give rise to loss of ammonia, so under these conditions the urine should be diluted with water. Smell can be reduced by aeration of the urine.

TABLE 3.6   *Analysis of urine*

Dry matter	5%
Organic matter C : N ratio	Insignificant amount of organic carbon
N	0.2%
P_2O_5	0.02%
K_2O	0.6%

Slurry

This semi-liquid product arises from modern housing methods where stock are kept on slatted floors. The product is a mixture of faeces and urine with possible dilution with washing-down water and bedding. The housing method would not be acceptable to appropriate farmers, but intensive livestock units with very little land on which to dispose of the slurry make it available to third parties. However, it has recently been shown that slurries derived from animals fed drugs may impede the composting process. Problems also arise in that the biggest build-up of the material is in winter when the use of heavy equipment on wet land can give rise to soil structure damage.

TABLE 3.7   *Analysis of various slurries*

	Cattle	Pig	Hen
Dry matter	10%	15%	80%
Nitrogen	0.4%	0.6%	1.0%
P_2O_5	0.25%	0.5%	0.75%
K_2O	0.5%	0.25%	0.5%
Organic matter	5.5%	6.0%	11.0%
Quantity per animal per day	50 kg	4 kg	0.5 kg

As slurries by definition are made up of urine, faeces and washing-down water, they contain very little straw or other bedding and can be considered, like urine, as a source of nitrogen to balance products such as straw or wood shavings which have a high carbon:nitrogen ratio. In general slurry causes smell pollution problems, and aeration will convert about half of the nitrogen to the ammonium form, which can give scorch problems on application. Although slurry should be aerated to avoid smell pollution, temperature will not rise sufficiently to kill pathogens and weed seeds. This is particularly important with pig slurry where dock seeds, imported through the feed grain, can be a very serious problem. Most readers will also be aware of the vigorous volunteer tomato plants seen where night soil has been used. Most of the nitrogen in chicken manure is already in the ammonium form, and not only will there be plant scorch and loss of ammonia but also the ammonium which will be added to the soil in comparatively heavy doses will cause imbalance in the organic clay soil domains. Therefore all chicken slurry and manure should be composted before use. Again with pig slurry the use of high levels of soluble copper salts in the pig finishing diets could give rise to high copper levels in the slurry, which in turn could raise the copper levels in the soil to unacceptable proportions. Copper in the soil is probably phytotoxic at 100 ppm.

Overuse of slurry can give rise to damage to soil structure, and the high nitrate content of some slurries would cause excesses to be leached from the soil. This is undesirable from a farming and environmental point of view, and would also interfere with the ethylene cycle. In general slurry applications should be kept below 60 cubic metres per hectare on permanent grass or older leys, whilst on arable land or young leys a quarter of this amount would be more appropriate.

Straw

The very high carbon:nitrogen ratio means that when straw is ploughed in after harvest, depletion of the nitrogen content of the soil will occur. This is because the micro-organisms which degrade the straw need a nitrogen component, and this they will take from the nitrogenous compounds in the soil. In soils already low in nitrogen, decomposition may take a great deal of time. The

TABLE 3.8 *Analysis of straw*

Dry matter	95%
Organic matter C : N ratio	
Oats	40 : 1
Wheat	100 : 1
Nitrogen	0.5%
P_2O_5	0.3%
K_2O	1.2%

formation of organic acids during this slow decomposition may affect germination of the following crop.

Professor Dhar[6] of the Sheela Dhar Institute of Soil Science, University of Allahabad, showed in the 1950s that the application of insoluble phosphates to paddy straw residues prevented reduction of nitrogen in the soil. He found that if insoluble phosphate fertilizers were added to the straw at a rate of 1% of straw yield, and the whole was then ploughed, not only was there no nitrogen loss but atmospheric nitrogen was fixed and this fixed nitrogen was available to subsequent crops.

On an acidic Kari soil an experiment was conducted by Dhar, with the following treatments:

1. 75 kg of nitrogen was applied to paddy stubble before the first ploughing;
2. 1% basic slag was applied to stubble before the first ploughing;
3. control.

These paddy crops were grown without further addition of manure or fertilizer. The results obtained are shown in Table 3.9. Similar results were obtained by Dhar on alkaline soils.

Plot experiments with similar treatments were carried out on farms in the U.K. in 1976: the cereal was barley. Under field conditions there was a 4.7% increase in yield over the control which had received in the autumn 40 units of phosphorus and potassium and a spring top dressing of 50 units of nitrogen.

However in spite of Dhar's work showing that straw can be incorporated into soil without denitrification occurring, the appropriate farmer would find the straw more valuable as bedding and subsequent composting of the farmyard manure.

TABLE 3.9

	1st Crop yield (kg)	2nd Crop yield (kg)	3rd crop yield (kg)
75 kg nitrogen	1,156	592	578
1% basic slag	1,526	918	859
Control	889	608	597

Compost

Properly made compost is a completely safe material which can be used at any rate. There is no possibility of causing scorch or other damage either to crops or to the soil structure. The only limitation will be the quantity which is available on the farm. Its use enhances the organic content of the soil.

Compost Making. Ideal composting is a process in which living organisms, utilizing oxygen in the air, feed upon the organic matter and from this material build up their own cell structures. All the nutrients, carbohydrates, nitrogen, phosphorus and other macro- and micro-nutrients are found in the organic matter.

Most of the carbon in the material for composting is in the form of carbohydrates, and the oxidation of this by the micro-organisms is in reality respiration, and so carbon dioxide and energy, which is dissipated as heat, are released. About two-thirds of the carbon present is required for the respiration, and the remainder is needed for recombination with the nitrogen in the material to form new microbial cell structure. A definite balance between carbon and nitrogen in the compost heap is required, and it is usual to try to develop the heap with an organic carbon : nitrogen ratio of between 25 : 1 and 30 : 1. It should be noted that the carbon referred to is that having its origin in "organic" sources (as opposed to carbonates). If there is an excess of carbon in the heap activity decreases and some of the organisms die, so allowing the nitrogen in their own body cells to become available as a nitrogen source for the remaining organisms. After several of these cycles the carbon : nitrogen ratio will become correct and final decomposition occurs. If, on the other hand, there is an excess of nitrogenous compound in the heap, then the organisms break down this excess into ammonia which is lost from the heap. Both of these conditions are undesirable as not only are they wasteful of potential nutrient but they can also delay the process.

The micro-organisms involved in the process are mainly bacteria and fungi. The mesophyllic bacteria act early in the composting process. They are aerobic in nature and will continue to work up to temperatures of about 50 °C. The thermophylic bacteria take over the process beyond this temperature and continue the composting process up to about 75 °C. The fungi found in plant material are also of value in breaking down material that was once alive. They

TABLE 3.10   *Composition of compost*

Dry matter	40%
Nitrogen	0.3%
P_2O_5	0.2%
K_2O	0.4%

TABLE 3.11

Organism	
Salmonella spp.	Death within 1 h at 55 °C and within 20 min at 60 °C
E. coli	Most die within 1 h at 55 °C and all within 20 min at 60 °C
Entamoeba	Death within a few minutes at 45 °C
Taenia	Death within a few minutes at 55 °C
Brucella spp.	Death within 3 min at 62–63 °C
Trichinella	Killed instantly at 60 °C
Streptococcus	Death within 10 min at 50 °C
Mycobacterium tuberculosis	Death within 15–20 min at 67 °C
Corynebacterium	Death within 45 min at 55 °C
Ascaris spp. eggs	Death in less than 60 min at 55 °C

are nature's great scavengers. During respiration heat is generated; the outer layers of the heap act as insulating material and so temperatures in excess of 70 °C are attained. At these temperatures, as Table 3.11 shows, pathogens are destroyed. Also at these temperatures weed seeds will be killed. However, as we have seen, the outer layers of the compost heap act as insulation layers and do not heat, and so it is essential to turn the heap to ensure that all pathogens are killed.

In practice it is therefore important to take steps to ensure that the materials to be composted are mixed to give a carbon:nitrogen ratio of about 30:1, so that nutrient loss does not occur, and that there is a plentiful supply of air to ensure the process is aerobic, so producing the essential temperature rise. The blend of materials must supply moisture, required by the organisms, at about 60%. The determination of the nitrogen content of a potential compostable material is relatively easy, and can be done by a local analyst. Carbon determination, on the other hand, is difficult. However, a good estimate of the carbon content can be obtained by use of the following formula:

$$\text{Percentage carbon is approximately equal to } \frac{100 - \text{percentage ash content}}{1.8}$$

The method of compost making developed at the University of California is ideal. In this method material for composting is chopped into pieces not exceeding 20 cm in length and mixed so that the final carbon:nitrogen ratio is about 25:1. The material is then heaped. The heap is triangular in cross-section, about 2 m wide at the base and $1\frac{1}{2}$ m high. The chopping and mixing allows access of plenty of air and so good aerobic fermentation takes place. The heap is turned after 3 weeks. In farming situations a rear-ended manure spreader is often used to make compost heaps. The manures and straws are loaded onto the spreader with a front end tractor-mounted fork; the spreader is kept stationary, being driven by a second power source. The material travels along the bed of the spreader and as it is ejected at the rear it is shredded and thrown into a heap of the correct cross-section. The despatch of the material at

FIG. 3.2A   Diagram of a University of California compost windrow. The materials to be composted are chopped and mixed so as to give a carbon:nitrogen level of about 25:1 and a moisture content of about 50%.

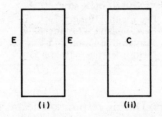

FIG. 3.2B   Plan of University of California compost heap. It is essential, when turning the heap from position (i) to position (ii), that the outside edges E in position (i) are placed in the centre of position (ii).

the end of the spreader ensures the inclusion of plenty of air. Once the heap has heated (usually after about 3 weeks) it is turned using the same technique but taking care that the outside of the old heap is brought to the centre of the new. Under very adverse conditions of cold or wet it is worthwhile trying to cover the heap to make sure that the temperature rise is obtained. Should the moisture level exceed 60% more frequent turning will be required. If the heap becomes very wet anaerobic fermentation is a real possibility, resulting in a wet slimy pile which, though decomposed, because there has been no temperature rise, can be full of pathogens and weed seeds. Where there is a great deal of compost being made it might be worthwhile investing in one of the machines which have been designed for the purpose of chopping, mixing and turning compost.

The Indore method of composting developed by Sir Albert Howard is named after the area of India where he carried out his original work. It is a method whereby the material to be composted is layered so that the layers alternate between low and high nitrogen content material. The pile is again built to a height of about $1\frac{1}{2}$ m and 2 m wide at the base, and the top will be about 1.3 m wide. The heap is covered with about 5 cm of soil, which deters flies and prevents the emission of foul odours which might occur during the process. The composting process takes about a year in temperate climates if no turning is practised. However in the hot Indian climate the heap was turned after 10 days and then turned again about a month later. The compost was usable about a month after the second turning. The disadvantage of the Indore method is that it is very easy to obtain anaerobic fermentation as the material was not well mixed in the first instance, and also because the heap is more square in cross-section. Of course if the process does go anaerobic then pathogens

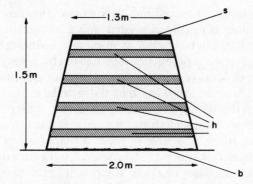

FIG. 3.3   Cross-section of an "Indore" compost heap. s: Layer of soil to seal the heap; h: material with high N content (such as manure) spread in layers between material with high C content (such as straw); b: the heap is often built on a brushwood base to allow access of air.

survive, and in countries where night soil is often used as the nitrogen source this can be a human health hazard.

As has already been discussed, so as to reduce nutrient loss to the farm, sewage and garbage should be recovered from the urban areas. Aerobic composting is the ideal process for this return route to be made. The high temperature will kill all the human pathogens and turn potentially hazardous material to a safe fertilizer and soil conditioner. However, the problem of heavy metal contamination has to be accepted, and either until economic methods to remove the phytotoxic materials are discovered, or until such time as industry is encouraged and people are re-educated to separate their refuse, this source of nutrient is not available to the farmer. The word re-educated is used because it is recorded by E. J. Simmons in his book *Memoires of a Station Master*, which was published in 1879, that a signalman stole garbage from wagons carrying London refuse which had been sold for use as manures by farmers. Of course in 1879 there was no admixture in the refuse of undegradable plastic or tin cans. More recently in the Second World War most householders realized the importance of separating their refuse into that which was suitable for recycling (cans and paper), that which could be fed as pig swill or for fertilizer, and bones for processing into glue and insoluble phosphate fertilizers.

Contaminated garbage, composted and sieved to remove tins and undegradable plastic, could if not severely polluted be available to the forester, where one application for every crop of trees, say every 50 years, would not cause a serious build-up of toxic metals.

Methane Compost Residue

As already noted, decomposition of organic refuse can take place anaerobically, and this is the common putrefaction seen in swamps. The carbon not used by the micro-organisms is turned into "marsh gas" (methane). The process in natural conditions is usually associated with a foul smell, but under proper control the process can be of value as an energy-producing source. Anaerobic composting for methane production is carried out by making sure that the heap is well compressed, to keep out the air, and kept at a moisture content in excess of 60%–70–75% seems ideal. In China the material is placed in cement-lined pits and the gas is collected through a pipe at the top of the pit. Frequently most of the obnoxious odour in methane production is due to hydrogen sulphide, a poisonous gas, which is soluble in water. Composting in a pit with a high moisture content allows the gas to be dissolved and a great deal of the nuisance is removed. The methane produced is piped to a gasholder, whence it can be used as an energy source around the farm. In theory 1 tonne of waste should produce about 60 cubic metres of methane, but at a temperature of 15 °C this will take about a year to gather. However, if the temperature of the digester is raised to 30°C the digestion period is reduced to about 6 weeks, so in winter in temperate zones some of the methane produced is used to heat the digesters. In conclusion it must be restated that where night soil has been used in methane production, the resulting compost should not be used on crops destined for human consumption. Indeed it is doubtful, because of this and the weed seed nuisance, whether compost from methane produced with human night soil has a place in appropriate agriculture. Of course it may be possible to recompost this aerobically but the carbon : nitrogen ratio would need to be recalculated. The best return of waste organic material is through the aerobic route.

Non-farm Origin

Blood Meal

Analysis – dry matter 95%; nitrogen 12%.
   A slow-acting expensive slaughterhouse waste more often used as a protein source.

Bone Meal

Analysis – dry matter 95%; P_2O_5 30%.
   A slow-acting phosphorus fertilizer.

Guano

Analysis – dry matter 95%; nitrogen 12%; P_2O_5 6%; K_2O 3.5%.
   A material derived from the accumulation of many years of seabird droppings which was once extensively mined on islands off the west coast of South America. Its use is mainly confined to horticulture. (From the holistic point of view it is noteworthy that the development of very large-scale anchovy fishing

to produce fish meal fertilizers not only greatly diminished the stock of anchovies for this purpose but also largely removed the food supply of the guano birds, reducing their numbers catastrophically. The danger now is that both guano and anchovy stocks will become exhausted.)

Hoof and Horn

Analysis – dry matter 95%; nitrogen 9%; P_2O_5 35%.

A very expensive slow-release nitrogen–phosphorus manure used almost exclusively in horticulture.

3.5 Mineral Fertilizers

Commercially made soluble fertilizers, such as those based on ammonium nitrate, are not acceptable in appropriate agriculture. Their upsetting of the nutrient balances in the soil, especially those surrounding the clay domains, is bad enough. The luxuriant growth they produce, being more likely to be attacked by fungi and insect pests, especially aphids, which can be carriers for virus infection, leads the farmer to apply yet more commercially made chemicals in the form of fungicides, pesticides, etc. However, many of the raw materials used in the manufacture of soluble fertilizers can themselves find a place in appropriate agriculture as they are insoluble and are only slowly released into the soil ecosystem.

Mined Rocks

The reliance of a practical agricultural system on mined sources for its fertilizer is not, of course, sustainable, and their use should be only one of last resort.

Rock Phosphates

Phosphate rocks are mined in various parts of the world but most of western European supplies come from Tunisia or Senegal. The Tunisian phosphates are mainly composed of calcium salts and are most effective in acid soils. Those from Senegalese sources contain much higher proportions of aluminium and should be used on soils which are neutral or alkaline because under acid conditions aluminium can become phytotoxic.

Rock Potassium Minerals

Many feldspars and basalt rocks contain about 8% of potassium salts, and if these rocks are finely ground then the material can be used as a slow-release potassium fertilizer.

Calcified Seaweed

This product, taken from the seabed off the coasts of Cornwall and Brittany, contains very high levels of calcium and, perhaps more importantly, high levels of trace plant nutrients. The material is used as a liming source, albeit very expensive, and also as a source of trace minerals which on a well-organized appropriate farm should not be needed. As an aside it is not without interest that this material, rich in copper, is found naturally near those soils which are copper-deficient. Is this one of the amazing provisions of nature?

Chilean Nitrate

Chilean nitrate was the first fertilizer to be imported into the United Kingdom early in the nineteenth century. It is not a satisfactory fertilizer for the appropriate farmer because, although it contains the quickly absorbed nitrate ion which would be useful when plants are under stress conditions, it also contains about 25% sodium which will impair soil structure especially on clay soils, and the high nitrate content will tend to suppress the vital ethylene cycle. Again the appropriate farmer would consider very carefully the use of a mined resource which might be of more benefit to the farmers of its country of origin.

Industrial By-products

There are several well-defined by-products which are of potential value. The problem of heavy metal contamination of sewage and garbage has already been discussed, but heavy metals can also be contaminants of other industrial by-products which could find a place as fertilizers. The metal may be damaging in two ways:

1. The heavy metal may be capable of being taken up by the plant and the resultant crop may be toxic to either or both animals and man. The most toxic metals in this respect are cadmium and lead. (Mercury, molybdenum, arsenic, selenium, chromium and fluorine are direct poisons, and only become a threat if they are actually eaten by the animal.) It is generally agreed that the maximum quantity of cadmium which can *ever* be applied to land is 5 kg per hectare and for lead the maximum is 1000 kg. Not more than one-fifth of this quantity can be applied in *any 6-year period*.
2. Other heavy metals, copper, nickel and zinc, are toxic to plants. Once any heavy metal has been applied to land it is impossible, in the present state of knowledge, to render it harmless and further additions of metals just make the land more toxic. The British Ministry of Agriculture, Fisheries and Food, with the help of research, have shown that 560 kg of zinc per hectare can be applied to land with no adverse effect as long as it is applied over a

very long period – say 30 years. They have also shown that copper is twice as toxic as zinc, and nickel is four times as toxic as copper, and with this knowledge they have coined a term "zinc equivalent", which acts as a guide to the toxicity of the material. The zinc equivalent of copper is therefore 2 and of nickel 8. Take for example an industrial by-product rich in calcium and potassium which is potentially a valuable potash fertilizer. Analysis of the material shows that it contains, on a dry matter basis, 125 mg/kg of copper, 20 mg/kg of nickel and 175 mg/kg of zinc. The zinc equivalent of this material is: $(125 \times 2) + (20 \times 8) + (175 \times 1) = 585$ mg/kg of zinc equivalent. Therefore 1 tonne of this material contains 0.585 kg of zinc equivalent. As it is safe to apply 560 kg of zinc equivalent per hectare over a period of 30 years then a total of 560 divided by 0.585 tonnes, i.e. about 950 tonnes, could be spread over a hectare over a period of 30 years. As a precaution it is generally advised that no more than one-fifth of the permissible quantity is applied every 6 years. This type of product could be deemed to be safe as it is unlikely that anyone would wish to apply 190 tonnes at any one time. However, similar exercises carried out on sewage sludge obtained from heavy industrial areas can give zinc equivalents of over 7000 mg/kg, and the use of such products is not possible. It must be emphasized that in the state of present knowledge once the zinc equivalent level for the land has reached toxicity (over whatever period of time) then no further applications of heavy metals must ever be made.

Special regard must be applied for boron, not often found in sewage, which can be an industrial effluent. It is generally agreed that any product containing more than 50 mg/kg of water-soluble boron on a dry matter basis should not be used.

Basic Slag

Basic slag was a by-product of the pig iron industry and contains about 20–25% of P_2O_5, lime and trace elements. It is highly prized as a fertilizer as it is barely soluble and is a valuable source of phosphorus and lime. Unfortunately, due to the change in steel-making in the United Kingdom, the product is becoming less easy to obtain. It is still available in some other countries.

Fly Ash

After coal has been burnt in the coal-fired electrical generating stations, fly ash is the resulting product. There are many thousands of tonnes available every year, and as the product contains up to 30% of potassium it is a potential source of material to restore this nutrient which is so severely lost each year. However, it is an extremely alkaline product containing large quantities of aluminium and boron which are often at a phytotoxic level. If it has been allowed to weather in heaps on sites near generating stations, then its toxicity is reduced. Nevertheless at the present state of knowledge its use would be restricted to the acid soils which are generally far removed from generating stations.

Cement Kiln Dust

In certain of the cement-making factories flue gases condense in the emission chimneys and the resultant product is not unlike fly ash: it contains about 30% of calcium and 5% potassium. Unfortunately some dusts also contain low levels of heavy metals and before its use on land it should be analysed and its zinc equivalent calculated.

Sugar Beet Factory Process Waste

Some sugar beet factories produce a lime slurry which can be used as a liming agent. As it has a very high moisture content it is not economical to carry the product any great distance.

Manufactured Fertilizers

In general fertilizers from this source would not be acceptable due to their non-sustainability, but under very adverse soil, climate or growing conditions there are two products which must be considered as of potential value.

Calcium Cyanamide

Calcium cyanamide is both herbicidal and fungicidal. If this chemical is applied to cereal crops where broad-leaved weeds are a problem the cyanamide will "stick" to the broad leaves and kill these plants, whilst it will slide off the more erect cereal leaves. The product contains about 20% nitrogen, and once it reaches the soil it is broken down by soil bacteria first to urea and then via the ammonium state to nitrate, where it can be utilized by the crop. The late release of the nitrate gives the crop a boost after the removal of the weed. Additionally, because the product is initially a calcium salt there is a slight residual liming effect. Calcium cyanamide is corrosive, finely ground and unpleasant to use. It is toxic to human beings if inhaled after the operator has consumed large quantities of alcohol. Because of its unpleasant nature it is often blended with 1% of oil to reduce the dustiness and to make the product more "sticky". In terms of appropriate agriculture cyanamide would not be used routinely, but it is a useful product where weeds threaten to destroy a crop or where fungal attack is out of control. (One immediately asks why have these situations occurred and how do I prevent them arising in the future?) Its use can be justified as it is not persistent in the soil, does not destroy soil microflora and, because it has to undergo change by biological activity before being taken up by the plant, it cannot severely upset the nutrient balance on the organic clay complexes.

Urea

Urea is a hygroscopic chemical and is usually mixed with diatomaceous earth before granulating. It contains about 42% of nitrogen. Urea is of course found in the urine of mammals; it is the method that the body has developed to get rid of unwanted nitrogen. It is therefore a very "organic" product, although it is now made in large quantities by industry. Once the urea reaches the soil it is changed by soil micro-organisms in the presence of water to ammonium carbonate, according to the following reaction:

$$CO(NH_2)_2 + H_2O \rightarrow (NH_4)_2CO_3$$

The ammonium carbonate is unstable and the ammonium ions are converted by other organisms in the presence of oxygen into the nitrate ion which is then available to the plant. The product therefore undergoes change by biological means before being taken up by plant roots. It can also be absorbed directly by plant leaves. Scorching can therefore occur under very warm conditions, and if large quantities of urea are applied to the soil then the balance is upset and ammonia is lost; the soil rejects that which it cannot deal with. Urea occurs at a level of about 3% in urine, and it is the urea that gives urine its status as a nitrogenous fertilizer. Clearly then it is permissible for the appropriate farmer to use urea as a nitrogenous fertilizer. However, its use without caution would give rise to all the problems – other than soil structure – which occur when soluble fertilizers based on ammonium nitrate are used. Additionally, urea manufacture uses a great deal of oil energy and is therefore not a fertilizer which will make for a sustainable agriculture. The advice concerning the use of urea then would be to avoid it except where there is dire economic necessity, such as very cold weather following a very wet winter when cereal tillering will fail. Under these conditions when applied in a 3% solution (as in urine), at a level to give about 30 kg of nitrogen per hectare, its use would be preferable to Chilean nitrate with all its ill-effects on soil structure and disturbance of the soil nutrient balance.

3.6 Plant Nutrient Loss

In wet tropical and temperate soils there is a continuous loss of potential plant nutrient by leaching. Indeed in tropical conditions the nitrogen supply is so rapidly lost from the soils that the reserve of nitrogen in that ecosystem is held in the leaves which continually drop, and which are rapidly converted by soil bacteria in the high temperatures into nitrate which is taken up by the plants. The felling of these forests removes the ecosystem's nitrogen supply, and before organic status can be built up in the soils, all plant life reduces and the area turns to desert.

Workers in Holland have spent several years measuring the losses of nutrient from bare soil in temperate conditions. Their work shows there is an annual leaching loss per hectare of 45 kg of potassium and 60 kg of nitrogen. There is little or no loss of phosphorus because of fixation. They have further calculated that the losses which can occur from the volatilization of nitrogen and dentrification of nitrate could amount to a further 70 kg of nitrogen. Thus the total losses of about 130 kg of nitrogen could be more than many farmers actually apply each year to crops. It is therefore essential that these severe losses are reduced to as low a level as is possible. This is best done by ensuring that the ground is never left in a bare state. The husbandry practice of keeping land covered between commercial crops is known as green manuring.

Green Manuring

The various biological systems which operate in the soil are continually making available plant nutrient, some of which will be used by the plant crop or weed. Once the crop is removed at harvest time, plant nutrients will move down the soil profile and will eventually be removed from the reach of plants. Green manuring is designed to arrest this loss and at the same time produce large quantities of green material which can be removed for composting or incorporated into the soil by ploughing. Ploughing-in is the least satisfactory as decomposition of the material may initially give rise to nitrogen loss as the carbon : nitrogen ratio of the material is unlikely to be near the desirable ratio of 25 or 30 to 1. Care must be exercised in the type of crop which is to be used as a green manure; it is very easy to slip into monocultural systems. For instance legumes, because they can fix nitrogen, are potentially the most valuable green manure, but if they were always sown with or after the crop it is possible that they would enable fungus disease specific to legumes to persist on the farm. The same would be true of the use of mustard between two brassica crops. Another precaution which must be observed is to ensure that the green manure does not become a weed. This should not occur as the green manure must not be allowed to seed, but in practical farming "the best laid plans of mice and men . . .". So that the nutrient loss is minimized, only rapidly growing species, particularly those which are deep-rooted, should be used. The deep-rooting allows of nutrient recovery at deeper levels, as well as helping to aerate soils at these levels. Cheapness of seed is also worthy of consideration.

The most common plants to be used are:

Legumes: trefoil and red clover

The legumes could act as autumn grazing or for silage making, whilst if they were grown between cereal crops the straw from the combine could be chopped and the nitrogen fixed by the legume would be sufficient to ensure rapid decomposition of the straw.

Brassicas: white mustard, stubble turnips and rape

These are all very rapidly growing plants and can be used as a crop after, say, early potatoes or cereals. Rape and stubble turnips give valuable grazing for livestock.

Grasses: Italian rye grass

This is a cheap aggressive plant and would not usually be sown under cereals. Its more usual place is in short-term leys where land is taken out of rotation for between 3 and 6 months. The crop could be sown with red clover and used firstly for winter roughage production followed by hard grazing.

Rumex: buckwheat

A very useful green manure of a botanical species not extensively cultivated. However, like other members of the family (docks and sorrel) it can become a severe weed of arable land.

Phacelia tanecifolia

This is not grown to any extent in the U.K. It has, however, all the virtues for use as a green manure. It belongs to a botanical family no member of which is grown under agricultural conditions. It produces a vast quantity of green bulk and, being extremely frost-susceptible, will never become a weed under U.K. conditions.

Mananagement of Green Manures

The practice of spreading uncomposted farmyard manure or surplus slurry onto a green manure crop is known as sheet composting. The nutrients from the applied manures will be taken up, and so a much heavier crop will result. Under these conditions stock could be allowed to graze the green manure, although all the previously discussed problems of plant and animal pathogens would have to be considered. More usually the bulk produced would be cut and composted, so providing safe material for use at a later date. If this latter course is adopted it is usual to wilt the crop after cutting, so obtaining a drop in the water content of the plant. Before composting, the carbon:nitrogen ratio would need to be adjusted, as would the moisture level. If the soil structure is not good then problems can occur if the green manure is ploughed in. Under these adverse conditions the lack of soil oxygen can give rise to the formation of organic acids, particularly acetic, in the soil. The presence of the acid will impair germination of future crops. If there is any doubt about soil structure then the crop should be composted, or alternatively at least 14 days allowed to elapse before ploughing in and sowing the next crop. This time lapse allows for some decomposition to take place before turning in.

The ultimate in the green manure approach occurs in the permanent grass-land of the prairie or in woodland ecosystems. In these conditions nutrient conservation is at its highest, and if the ecosystem is sufficiently varied then the permanence of the ecosystem does not appear to give rise to severe disease or insect pest problems. The reader has only to consider the growth of the massive trees in an oak/ash ecosystem and the nutrient needs of the aged plant to realize that the ecosystem can supply all the nutrient required without recourse to man's application of fertilizer. It is not an answer to say that the loss of nutrient from these systems is low; one has only to consider the old system of coppicing or the continuous cutting of the woods at Bradfield in Suffolk to produce oak for the maintenance of a local cathedral to realize that the statement is untrue. It is therefore very desirable to attempt to develop ecosystems for differing climates and topography so that ground cover could be maintained and yet commercial crops can be included. The use of the Wallace Soil Reconditioning Units give hope for great advance in this technique. Research is required to find the correct ground cover plants. In all probability these will be based on non-aggressive clovers and grasses such as birds-foot trefoil (which does not cause bloat problems) and meadow fescue which is not too aggressive. The use of trees, as defined in the Keyline systems discussed in Chapter 2, could be native species for food or timber. Permaculture as developed initially by Bill Mollison in New Zealand and Australia has shown the way ahead. All that is now needed is for each individual farmer to study the ultimate ecosystem for his area and then to devise methods by which he can grow his crops within that ecosystem. When that has been carried out the perfect balance of ecological agriculture will have been obtained.

References

1. Sources of sales figures: (a) Cereals – Home Grown Cereals Authority; (b) Livestock – Meat and Livestock Commission; (c) Milk – Milk Marketing Board; (d) Eggs – Eggs Authority.
2. Statistical Office, Ministry of Agriculture Fisheries and Food.
3. Waste Collection Statistics 1984. Chartered Institute of Public Finance and Accountancy, London.
4. Higginson, A. E. (1982) *The Analysis of Domestic Refuse*. Institute of Waste Management Publication No. 10.
5. University of California Sanitary Engineering Technical Bulletin.
6. Dhar, N. R. (1986) Value of organic matter, phosphates and sunlight in nitrogen fixation and fertility improvement in world soils. *Pontificiae Acad. Scient. Scripta Varia*, 32, 243–360.

Chapter 4

Pests and Diseases

In any natural plant association all the species have come together either because they can all live together under the climate and soil topography, needing the same nutrients which are in plentiful supply, or because one species is advantageous to another. For instance the growth of a leguminous plant in a grassland ensures that the grass receives from the rhizobium, living symbiotically on the legume's roots, simple nitrogenous compounds which become available for the growth of the grass species. In this particular case the plant association has widened even further to include bacteria which are classified as plants, albeit they have no chlorophyll and cannot make the essential foods from carbon dioxide and water. In this case they obtain their nutrients from the leguminous plant.

As soon as man decides to introduce into the natural plant association other plants, which he calls crops, then he has upset the natural balance and it is his husbandry which will enable him to bring his sown plant to yield seed or roots. Other plants which he has not sown, but which will thrive in the new association, he calls weeds. It is the husbandry exercised by the farmer which limits or eradicates the naturally occurring plant species. It is possible, and indeed often happens, that the introduction of the crop so disturbs the natural plant association that not only do the weed species change but the fungi and insects which occurred in the original plant association with little or no effect now become extremely difficult to control.

4.1 Weeds

Modern crop husbandry is obsessionally directed at the control of weeds – at the least to their being minimized and at the best totally eradicated. This is because the weed in excess will deprive the crop of food, light and moisture and will take up space which could be used by the crop. Some weeds may be poisonous to farm livestock, may taint milk or, perhaps most importantly and commonly, act as an alternative host for fungus diseases and insect pests. An understanding of the environments in which plants grow is essential to the designing of methods to control weeds.

In farming there are only two types of plant community which are of importance – the open community (plants of bare land), of arable land, and the

61

closed community (plants of covered land) of grassland. However, in permanent pasture bare land will often occur where animal traffic is high, for instance around water troughs, gateways and along footpaths used by man. The grass is trampled out and the rosette plants, such as shepherd's purse, will become established. When weeds become a problem in any community they are most easily controlled by altering the community. For instance, wild oats will expand their grip in a cereal-growing monoculture, because the wild oat sheds its seed onto the ground about a fortnight to 3 weeks before the cereal is gathered, and so is ready to germinate in the next cereal seed bed. If, however, the crop were changed from cereal to, say, grassland the oat would eventually die out because firstly it can be prevented from seeding by mowing before it seeds and secondly the wild oat cannot survive intensive grazing. From the foregoing it does not need much deduction to realize that the changing of the environment or plant association frequently will reduce the presence of weeds. This is because the times of cultivation or harvesting will change, and the specialized conditions for the rapid spread of the weed alter annually. Table 4.1 shows how the incidence of weed seeds reduces dramatically as the crop is varied. The introduction of a different species is highly beneficial. If the environment is changed even further from arable to a closed grassland community then there is an even more dramatic effect on arable weeds.

All weeds can be classified into annual, biennial or perennial plants, and the method of eradication will depend on a knowledge of the life span of the weed. Annual weeds complete their life cycle in 1 year, and they are easily eradicated by ensuring that they do not flower and so produce seed. If the life cycle is thus interrupted the plant dies at the end of the year, and as it has not seeded it will be no further problem. Biennial weeds store food usually in tap roots in their first year, flower in the second year and then die. Again eradication is simple by ensuring they do not flower and produce seed.

Perennial weeds, however, being longer-lived, are not controlled by stopping seed production which only prevents the weed being spread to new sites. In general annual and biennial weeds are most troublesome in arable land where it is difficult, by the very nature of the crop, to stop the weeds flowering. Perennial weeds are most troublesome in grassland where, although it is easy to stop them seeding, it is difficult to eradicate the plants by other cultural methods. Fortunately many of the perennial plants do not stand grazing, and are eradicated by subjecting the ley either to intensive grazing by both cattle and sheep or to continuous mowing.

TABLE 4.1   Incidence of weed seeds
(per square metre)

Wheat monoculture	525
Wheat/wheat/maize	224
Wheat/maize	183
Wheat/maize/soya	60

The different grazing methods of cattle and sheep can tackle the weeds irrespective of their size. The habit of growth of weeds is also an important factor in weed control. Those such as fat-hen, which have a simple erect habit, may be easily mown even in an arable crop, and seeding can thus be reduced. On the other hand annuals such as chickweed and speedwell, with a procumbent habit, cannot be mown but are easily controlled by changing the association to grassland where there is no bare ground into which they can seed. Plants such as couch, which have rhizomatous growth, are extremely difficult to eradicate. Should they be present in arable land then harrowing or bare fallow will reduce their incidence, but again turning the field to grass for a period of 3–4 years will kill out the weed because it cannot tolerate grazing. Plants such as creeping buttercup, which are perennial with a stoloniferous habit of growth, are not easily eradicated. Grazing tends to encourage the plant to produce more stolons. If weeds of this nature are a problem in grassland then it is best to break out of grass or to implement the severe cultivations of fallow. Docks will not survive their crowns being eaten out by sheep, and so very intense grazing will eventually eliminate them.

So far only the common cultural methods for the control of weeds have been considered, but other cultural methods should not be forgotten. For instance in damp environments drainage will eradicate rushes or other water plants. The application of plant nutrients may often stimulate the crop so that it can compete effectively against the weed, as in the spreading of lime on acid soils which will encourage the growth of the lime-loving plants but be harmful to the acid-lovers such as sheep's sorrel, spurrey or heath bedstraw. It will be appreciated that these cultural methods will never wholly eradicate weeds, so that beneficial life in the environment is not destroyed.

In the past 35 years the agrochemical industry has been able to produce chemicals which are species-specific in that they kill off certain plants whilst leaving others unharmed. This technique, however, is not acceptable to the appropriate agriculturalist. The injection of relatively large quantities of persistent, highly toxic albeit specific chemicals into the environment will, in the first instance, leave a space which can be filled by an equally noxious pest (it will certainly not be filled by the crop, which is established by the time the agrochemical is applied to protect it). It is a fact that, as nature abhors a vacuum, so a growing environment abhors a space. If part of the environment is made vacant by the use of persistent chemicals then species or genera not killed by the chemical will increase to fill the vacated space. If it so happens that the space is filled by a resistant strain of the species originally attacked then the chemical farmer has a real problem. His answer is to use a different and generally more powerful chemical. The whole point that any space *will* be filled by some growing thing has been missed; the latest agrochemical practice of using a "cocktail" only delays the onset of resistance. Indeed it could be argued that when resistance does arrive, because known chemicals have been used, there will not be a suitable "killing" agent immediately available. This is not so far-fetched as it may seem; the medical and veterinary profession now reserve to their own profession antibiotics to deal with infection to ensure that human disease bacteria do not develop resistance. Resistance has in the past been commonly associated with bacteria and fungi, but has now been identified in plants. Resistant varieties of fat-hen have become prevalent in some parts of the world and eradication of them now has to rely on appropriate cultural techniques.

Allelopathy

It is now understood that certain plants produce chemicals which do not persist in the environment into which they are released, but when released by the donor plant can act as herbicides or pesticides. The production of chemicals by plants, and their effect on other species or even genera, is known as allelopathy – a branch of pest control now under active consideration by many scientists but whose origins lie in the myths and legends which have become folklore and are carried on by the cottage gardener of today as companion planting. Companion planting is basically only interested in the enhancement of a crop by another crop (a well-known example is the growing of garlic amongst roses to control aphis, and is widely claimed to improve the scent of the rose!); allelopathology is also interested in the lethal or inhibiting effect of one plant on another. Allelochemicals can be produced by any part of the plant (even pollen of some plants has killing properties), but mainly by roots and leaves.

The roots are of particular interest. A nematode soil worm (not earthworm) – *Trichodorus* – multiplies very rapidly on tomato roots and will kill the plant. However, asparagus roots exude a chemical which kills *Trichodorus* and it is so effective that even a mixed planting of asparagus and tomato controls and will eventually eliminate *Trichodorus*. Again there are chillis which are resistant to attack by the *Fusarium* fungus. It is now established that these varieties give off exudates from their roots which eliminate or inhibit germination of the pathogenic fungus. Exudates from the roots of non-resistant chilli plants have no effect on the fungus. It is therefore logical to believe that in this case the resistance is dependent not on the actual structure of the plant but in its ability to manufacture toxins.

There exist three interactions which have to be considered:

1. Crop plant with crop plant

(a) Barley undersown with grass and clover will give a poor take of the subsidiary crop because barley roots exude a chemical inhibiting to the germination of grass and clover seed.[1]

(b) Clover sickness: the roots of red clover exude a chemical which will impede the germination of its own seed at 100 ppm; at 300 ppm it will have a similar effect on white clover; at 700 ppm on alsike and at 1000 ppm on vetches.[2]

(c) It is well established that when rice waste is spread onto a paddy field the second crop is not as good as the first.[3]

(d) In contrast, exudates from pea and vetch roots stimulate absorption and uptake of phosphorus, nitrogen, potassium and calcium. It has been established that this enhancement is not due to the rhizobial nitrogen release.[4]

(e) Lupin and mustard stimulate the growth of buckwheat. Contrarily, established buckwheat hampers lupin germination.[5]

2. Weed Plant against Crop Plant

(a) Many members of the heather (*Erica*) family hinder growth of other crops.[6]

(b) *Rumex* (dock) leaves, when decaying, deter germination of seeds of other species.[7]

(c) Couch grass (*Agropyron*) reduces wheat yields. This is due to the blocking of phosphate uptake, and for this reason application of phosphate fertilizers will not overcome the deficiency.[8,9]

(d) The presence of annual Californian thistle curbs seed germination.[10]

(e) It is difficult in practice to establish good grass varieties into pastures where native grasses exist, and this is believed to be due to the allelopathic effect of the native species.[11]

(f) Wheat yields are appreciably increased when growing in mixed stands with corn cockle (*Agrostemma githago*).[12,13]

3. Crop Plant against Weed Plant

(a) Lupin species restrain the growth of fat-hen.[14]

(b) It has been noted that natural fescue meadows in Kentucky do not have weed, and this, it is believed, is due to exudates from the roots of the fescue repressing germination of other species.[15]

(c) In the screening of over 3000 cultivars of *Avena* (oat) it has been found that about 25 have the potential to release, by way of their roots into the soil environment, a chemical which is lethal to all cruciferae species.[16]

Allelopathy can be used in appropriate agriculture as a method of controlling the effects of pests on plants and, maybe in the future, on animals. It must be realized that the allelopathy involves the release of chemicals into the ecosystem, and it is these chemicals which have their beneficial or harmful effect on the crop. The important thing is that all allelochemicals, once released, are short-lived in the environment and therefore do not disastrously upset the balance as persistent chemicals would. The application of persistent chemicals into an environment upsets the balance to such an extent that genetically resistant species, whether they be bacteria, fungus or plant, become dominant in the ecosystem and are then generally uncontrollable. In the present state of knowledge allelopathy shows the greatest promise in:

1. The use of oats instead of barley as a cover crop for establishing grass ley.
2. The sowing of 1–2 kg/ha of white mustard in cereal crop, especially wheat, to increase yield because members of the Brassica family, especially white mustard, release from their roots oils which when broken down release hydrogen sulphide. This is more lethal than hydrogen cyanide and will sterilize the micro-environment and allow other crops to grow without interruption.
3. The sowing of one of the known oat anti-cruciferae varieties where cruciferous weeds are a problem. This indicates the possibility of breeding plants to exploit not only yield but also the presence of allelopathic chemicals, which may be exuded either to be of benefit to other plants or to impede the undesirable plants or pests.
4. Corncockle and vetch, when grown together with wheat without the use of added fertilizers, can produce yields of up to 4 tonnes/hectare of wheat. The response is more than that which would be expected from the release of nitrogen by rhizobial activity on the vetch.

This area is receiving bracken allelochemicals.
It is newly sown pasture looking very sick
when compared to the rest of the pasture.

PLATE 5. *Allelopathic effects of bracken.* Although bracken has been cleared from newly sown pasture, the ley (enclosed by dark line) has not established due to allelochemicals from bracken still present in wood in background.

4.2 Pests

There is what can only be described as a multitude of invertebrates which can adversely affect the performance of both crops and animals, and it is not within the scope of this book to list each one with its specific method of control or eradication. However, in any agricultural system the presence of pests has to be acknowledged, and a broad outline of how they can be controlled is important. The conventional farmer will generally tackle the problems in a reductionist manner, by attempting to kill the pest outright, but experience has shown that this rarely succeeds in the long term and "resistant" strains appear. The holistic farmer will, on the other hand, consider the problem as a whole, and will adjust his farming methods so that a situation will develop where the pest is kept under control.

The best defence is obtained by the use of rotations, and once again it cannot be over-emphasized that by the simple act of a change of crop (animal or plant) the potential pest finds itself in a less hospitable environment, and does not reproduce so rapidly that the increase in numbers threatens the performance of the plant or animal.

An illustration of a monocultural parasite build-up is the "calf paddock", traditionally a piece of permanent grass close to the farm buildings where every spring the winter's calves were turned out to graze the lush young grass. The larvae of the husk worm, having overwintered in the soil, climbed to the tips of the young grass where they were eaten, completed their life cycle in the calf and the eggs were returned to the paddock from the newly infected calves. Each year there were more larvae, and so the infection became more severe. The case of husk in cattle is of interest, because it illustrates that a monoculture of stock can be as disatrous as a crop monoculture.

Most of the invertebrate pests have a complex life cycle and the larvae will commonly have a secondary host. If the husbandry method adopted can break the life cycle of the pest then their numbers will be controlled so that any damage they cause will not be of economic importance. Consider for example the serious sheep and cattle pest, liver fluke. The fluke eggs pass by way of the bile duct into the digestive tract and out of the animal's system in the droppings. If moisture is present the eggs hatch and the larvae swim until they can infect the fresh-water snail (*Limnaea*). If within 2 days the larva does not find a *Limnaea* snail then it dies. The larva, once in the snail, develop and when the snail dies the changed larva then swim on to a blade of grass where it encysts, and in this encysted state it can stay alive for many years. Once the cyst is eaten by a cow or sheep the cyst "hatches out" and the consuming animal is infected. The last stage travels from the digestive tract to the animal's liver where it lives parasitically, passing out its eggs to start once more a further snail infection. The fluke can be controlled in two ways. Firstly by removing the snail, and ducks will do this very effectively; and secondly, because moisture is essential twice in the life cycle, by drainage of the land. Fencing of the wet area will at least hinder, if not stop, the liver fluke's life cycle.

Practical farm hygiene also reduces pest problems. For instance tapeworm of the dog and fox has the sheep as its secondary host. In this stage it is known as a bladder worm and lodges in the sheep's brain, causing "gib" or "staggers". Obviously a dog or fox given an uncooked infected sheep's head will become infected in turn, and so allow completion of the tapeworm's life cycle.

In the plant kingdom pests can also, of course, be a great problem, and it is incumbent on the holistic farmer to adopt cultural methods which are harmful to the pest. Leather-jackets – the larval stage of the crane fly (daddy long-legs) – live on grass roots and do a great deal of harm. They pupate in July, the adult fly emerging in September. The fly only lives for about 4 days, but in this time it lays about 300 eggs on long grass. The eggs hatch, the larvae pass down the grass stems and live in the soil, feeding on grass roots until the following July. Therefore where permanent grassland is to be ploughed up it is best carried out in July, so burying the pupae. (Ploughing in August will give bare ground which will attract the wheat bulb fly). Even better control would be obtained if, in the year *before* ploughing, the grass were kept grazed very short in September, so depriving the fly of the environment she most most likes for egg-laying. The leather-jacket numbers would therefore be reduced before cultivations start.

The resistant cultivars, which give so much protection where fungus and virus are prevalent, are not so readily available as a method of invertebrate pest control, but it is worthwhile considering allelopathy as an aid. Differing plants do have differing abilities to resist pests. For instance, wireworm (the larva of the click beetle) is no real problem in grassland, but once the land is ploughed and cropped then the wireworm can be a very serious pest of cereals, potatoes and sugar beet. However linseed, mustard and rape are resistant to attack and peas, beans and kale are only slightly damaged. If, therefore, a wireworm count reveals a heavy infestation it is wise to plant a resistant crop. Whilst that crop is being grown, some wireworm will change via the pupa to the adult state and the population of wireworm will reduce because the adults will not lay eggs except on grassland. Once the wireworm count has been reduced the more susceptible crops can be grown without suffering economic damage.

4.3 Plant Diseases

Plants can be affected by four differing classes of disease – those attributable to:

1. deficiency of a trace element – grey speck in oats which is due to manganese deficiency;
2. bacteria – black leg in potatoes;
3. fungus – such as the rusts and mildews;
4. virus – the mosaic diseases of beet and potatoes.

Chemical Deficiency

Trace element deficiency will rarely if ever be seen in appropriate farming because steps are continuously taken to ensure that the delicate nutrient balances in the soil are not upset by the addition at any one time of large quantities of cation.

Bacterial Diseases

These are very rare in any type of crop husbandry, and they only become a problem where monoculture is practised, as it is only in these conditions where the bacteria can find the same suitable host every year, so allowing easy proliferation. When bacterial diseases strike they can only be overcome by a complete rest of the land (for many years) from the susceptible crop. Soil sterilization, chemical or heat, will kill not only the troublesome bacteria but also all the beneficial soil micro-organisms.

Fungus Diseases

Possibly this class of diseases poses the greatest threat to world agriculture. The use of persistent fungicides by the conventional farmer has led to the advance of strains of fungus which are resistant to the presently known chemicals, and it is proving increasingly difficult to find new chemicals which do give control. Plant breeders are turning their attention to the growing of genetically resistant varieties. Until such time as these new cultivars become available the growing of mixed varieties of a species is of help, as the density of susceptible plants in a field is reduced. In appropriate agriculture under farm conditions it is usual to try to avoid these diseases by taking preventive measures and so stop the disease from becoming established. Usually the soil is the most common site of infection, but infection can also come from manures which have not been properly composted, or from infected seed, or by the fungus over-wintering on other crops or weeds. Indeed the fight against plant disease is perhaps the most important aspect of the use of rotations. By altering the crop it is unlikely that the disease can carry over for the long periods before a crop is resown on any particular piece of land. Winter barley is a typical example of a crop which will lead to the spread of barley fungus disease, as it will provide a host for the fungus between spring-sown crops of barley. Again great attention should be paid to what may be called general sanitation. It is important to ensure that diseased plants or parts of plants are properly destroyed. Infected potato tubers may be fed to pigs and the fungus disease could pass through the manure to infect another crop of potatoes. Proper boiling of the tubers before feeding would have broken that particular life cycle, and composting of the manure would have made doubly sure of eradication. The removal of weeds will also help to reduce fungus infections as certain weeds may act as an alternative host to commercial crops.

Virus Diseases

This class of disease is becoming more of a problem in conventional farming. Most of the common viruses are carried by aphids, and as they suck sap from leaves the virus enters the plant from the proboscis of the aphid, or the aphid picks up the virus from infected plant sap and is able to carry the infection to the next plant it visits for food. In conventional agriculture the use of excessive quantities of nitrogenous fertilizers gives rise to sappy plant growth, and it is not a matter of dispute that conventionally grown crops have high moisture contents. These higher water levels increase the size of plant cells with a consequential thinning of the cell walls, allowing more easy access for the proboscis of the aphid. Work at the Pye Research Centre at Haughley in Suffolk, England, showed that the presence of aphids was always significantly lower on crops grown organically than those grown chemically.[17] The best control for aphids and virus disease is therefore obtained by limiting nitrogen intake, so ensuring that the plants do not produce sappy growth. Besides these managerial preventative methods, it is beneficial to grow varieties which have been bred to give resistance to disease. This new work of the plant breeder will no doubt be of great value. However, caution must be expressed, as it is possible, as with resistance to insecticides, that certain less common fungi or virus diseases may become more frequent as the present more easily controlled plant diseases are eliminated. In the end good sanitation and rotational cropping will give the best control of all diseases.

4.4 Animal Diseases

Other than the parasitic conditions which have already been mentioned, and which are well controlled by good husbandry, there still remain the other more serious diseases which need the attention of a qualified veterinary surgeon. The holistic farmer would hope that the condition would be treated in a holistic rather than a reductionist manner, and in general vets trained in homeopathy (which is concerned with the whole) would be more understanding than the conventionally reductionist trained vet.

It is worth recording that holistic agriculture, by its very nature of depending on a greater balance, is more likely to keep its animals free of the worst diseases of the respiratory and digestive tracts which are so common in conventional agriculture. The metabolic diseases such as milk fever and ketosis are hardly ever seen, because the cattle will be properly fed and not forced to produce yields up to their maximum genetic potential. It is equally true that there are today many farmers farming ecologically because the problems of infertility in dairy cattle, which were so severe and uncontrollable conventionally, suddenly disappeared when appropriate methods were adopted.

4.5 Fungicides and Pesticides

However well the farmer may build up plant associations which inhibit or reduce the effects of fungi and insects, it is certain that at some time, due usually to peculiar climatic conditions, fungi or insects may become a serious problem and threaten the whole crop. Obviously the holistic agriculturalist is farming to produce food, and severe threats have to be controlled as rapidly as possible. Mild damage to crop or plant is usually only of aesthetic importance. If it is necessary to resort to the use of chemicals then it is essential that the minimum damage is done to the environment and is of such a nature that the balance is rapidly restored. It is only permissible to use insecticides and fungicides under the most severe infestation and the chemicals which would be used will be those which have little or no persistence in the environment. This is in direct contradiction to the conventional agriculturalist, who searches for and uses the most persistent chemicals available. Generally the insecticide or fungicide will be of plant origin and will, by its very nature, be easily and rapidly assimilated into the environment. Indeed these plant "poisons" are part of allelopathy. With fungus diseases there do not as yet appear to be plant-produced chemicals which can be used, although the disease resistance of certain cultivars to fungus is probably due to chemicals produced by the plant which prevent the fungi spore growing. In the present state of knowledge simple chemicals based on the elements copper and sulphur are most frequently used.

Fungicides

It has been known for many years that sulphur and copper compounds have very good fungicidal properties, and indeed both can be used by the appropriate agriculturalist as neither is very persistent. All copper salts are fungicidal but the cheapest salt, copper sulphate, being very soluble, is not often used as it does not easily adhere to plant leaves. The copper salts most frequently used are those based on the hydroxide or carbonate, both of which are insoluble.

Copper Hydroxide Fungicides

The most common is Bordeaux mixture, which takes its name from the region of France where it was first used to control mildew on the vines. It is made by dissolving 5 kg of copper sulphate in about 150 litres of water in a wooden or plastic container. Once the sulphate is dissolved 6 kg of slaked lime, $Ca(OH)_2$, is added to the solution, which is thoroughly mixed, and the solution is then diluted to 500 litres. Care must be taken to ensure that the slaked lime does not contain any large lumps, as these can cause blocking of the spraying nozzles. Bordeaux mixture tends to crystallize out, and it is therefore wise only to make up sufficient for each day's use.

Colloidal copper oxide, which is a mixture of copper hydroxide and copper oxide, is as effective as Bordeaux mixture, is more easily distributed through sprays but is more expensive than Bordeaux mixture.

Copper Carbonate Fungicides

Burgundy mixture, also named after the wine-growing region, is the best known of this class. It is made by dissolving 5 kg of copper sulphate in 250 litres of water in a wooden or plastic container, and a similar quantity of sodium carbonate in 250 litres of water. The sodium carbonate solution is then added to the copper sulphate solution. Burgundy mixture is more expensive to make than Bordeaux mixture, but it tends to spread more easily.

Cheshunt Compound

Fifty grams of copper sulphate are mixed together in a dry state with 300 g of ammonium carbonate and left for 24 h, and then dissolved in water at the rate of 25 g of mixture in 8 litres of water. The resultant mixture is extremely effective but it is very expensive, and is usually only used on high-value crops.

Sulphur Fungicides

Flowers of sulphur has been used as a fungicide but the best product is lime sulphur. This is made by boiling together 50 kg of sulphur with 25 kg of burnt lime (CaO) in 250 litres of water. Both the sulphur and lime pass into solution and a reddish-yellow liquid remains. The lime sulphur is an excellent fungicide for the appropriate farmer as it is highly effective, but is quite unstable in air and light, and so rapidly passes out of the environment.

Some plants, especially the cucumber family, are very susceptible to sulphur – sulphur-shy – but if the lime sulphur is used diluted, damage to crops is avoided. Concentrated lime sulphur can be used when trees are dormant to control over-wintering fungus spores.

In general sulphur products are best for the control of the powdery mildews, and copper compounds better control downy mildews. Sulphur has the advantage that it has some effect on the killing of some insect pests.

Copper and sulphur fungicides should never be mixed together as they inactivate each other. It is usual to use copper products on potatoes for the control of blight.

4.6 Pesticides

Insecticides

As with fungicides the appropriate farmer will use those insecticides which do not persist in the environment, and in any case will only use any insecticide when the crop is likely to be destroyed. The most commonly used products are:

Nicotine

This is a highly effective insecticide, but it must only be used with care. The active ingredients in nicotine sprays are alkaline; therefore when they are sprayed on to acid fruits such as apples an insoluble compound of nicotine and malic acid is formed, and this stays on the surface of the fruit, killing any insect pests which attack the fruit. Because of its semi-persistence its use on acid fruits should be avoided whenever possible and, when used, great care must be exercised to ensure that the crop is well washed with rain before marketing.

If, however, the plant to be sprayed does not have any acid present then spraying should be undertaken on a warm day when the nicotine will vaporize and kill any predators present on the crop at that time.

Pyrethrum

This extract from a chrysanthemum species is a very powerful insecticide, but its expense tends to keep its use to horticultural crops which have a high value. It is very unstable in light and therefore the insecticide does not persist. Modern-day scientists have been able to change the structure of pyrethrum to remove the light-sensitivity property and the resultant insecticides are known as pyrethroids. Many insects have become resistant to pyrethrum due to the fact that the pyrethroids, being persistent, have killed off the susceptible species whilst allowing the resistant varieties to multiply without the natural competition which had been offered by the susceptible species.

Derris

This insecticide is obtained from the root of a tropical plant and is a very effective contact insecticide, 40 ppm being lethal to insects. Unfortunately the active ingredient, rotenone, is highly poisonous to fish, and care must be taken that it is not used near waterways. Derris is insoluble and therefore it has to be used mixed with inactive earths when it can be spread dry or, if it is to be carried in water, wetteners and spreaders have to be added to the water.

Wetting Agents

So as to ensure that an insecticide or fungicide spreads evenly over the leaf or the insect it is often advisable to add a wetting or spreading agent. The usual product which is used is ordinary soap, which will soon be lost from the environment. However in hard-water regions it is not a very good product to use as the hard-water reacts with the soap to form calcium salts which are insoluble. In these areas soapless detergents can be used, but care must be taken to use only products which are biologically degradable.

Vermifuges

Oil of Chenopodium

In piggeries the roundworm of pig (*Ascaris* species) can be troublesome as the eggs can live for long periods in cracks between troughs and floors. If the roundworm does present a problem then the use of oil of Chenopodium as directed by a homeopathic vet is very effective.

Garlic

Garlic, although tending to taint eggs and milk, is a very good vermifuge for all livestock.

A section dealing with insecticides and fungicides cannot be ended without some consideration of vaccination. Attitudes differ in differing countries – for instance vaccination against foot and mouth disease is not allowed in the U.K., and the holistic farmer will be guided by the law of the land in which he lives. It does, however, seem appropriate that where disease has become endemic vaccination should be employed.

Finally it cannot be too strongly emphasized that if good hygiene, good rotations and good plant and animal nutrition are practised the adverse effects of pests and diseases become much less because the animal or crop will "grow away" from the infection. The problems of human disease are always worst in those countries where the young and old are undernourished in overcrowded surroundings.

References

1. Overland, L. (1966) The role of allelopathic substances in the 'smother crop' barley. *Am. J. Bot.*, 53, 423–32.
2. Chang, C. F. *et al.* (1969) Chemical studies on clover sickness. II: Biological functions of isoflavenoids. *Agric. Biol. Chem.*, 33, 398–408.
3. Chou, C. H. and Lin, H.-J. (1967) Autointoxication mechanisms of *Oryza sativa*. I: Phytotoxic effects of decomposing rice residues in soil. *J. Chem. Ecol.*, 2, 353–67.
4. Rakhteenko, I. N. *et al.* (1973) Effect of water soluble metabolites of a series of crops on some physiological processes. In: *Physiological–biochemical Basis of Plant Interactions in Phytocenoses*, vol. 4, pp. 23–6 (in Russian; English summary).
5. Zabyalyendzik (1973) Allelopathic interaction of buckwheat and its components through root excretions. *Vyestsi. Akad. Nauk, BSSR Syer Biyal*, 5, 31–4 (in Russian).
6. Ballester, *et al.* (1971) Estudio de sustancias de crecimiento aisladas de *Erica. L. Acta Ci. Compostelana*, 8, 79–84.
7. Einhellig, F. A. and Rasmussen, J. A. (1973) Allelopathic effects of Rumes crispus on Amaranthus retroflexus, grain sorghum and field corn. *Am. Midl. Naturalist*, 90, 79–86.
8. Welbank, P. J. (1963) Toxin production during decay of *Agropyron repens*. *Weed Res.*, 3, 205–14.
9. Minar, J. (1974) The effect of couch grass on the growth and mineral uptake of wheat. *Felia. Fac. sci. Nat. Univ. Purkynianae Brün.*
10. Bendall, G. M. (1975) The allelopathic activity of Californian thistle in Tasmania. *Weed Res.* 15, 77–81.

11. Scott, D. (1975) Allelopathic interactions of resident tussock grassland species on germination of oversown seed. *N.Z. Jl Exp. Agric.*, **3**, 135–42.
12. Gajic, D. (1966) Interction between wheat and corn cockle on brown soil. *J. Sci. Agric. Res.*, **19**, 63–96.
13. Gajic, D. and Nicocevic, G. (1973) Chemical allelopathic effect of *Agrostemma githago* upon wheat. *Frag. Herb. Jugoslavia*, **18**, 1–5.
14. Dzubenko, N. N. and Ptrenko, N. I. (1971) On biochemical interaction of cultivated plant and weeds. In: *Physiological–biochemical Basis of Plant Interactions in Phytocenoses*, vol. 2, pp. 60–6 (in Russian; English summary).
15. Peters, E. J. (1968) Toxicity of tall fescue to rape and birdsfoot trefoil and seedlings. *Crop Sci.*, **8**, 650–3.
16. Fay, P. K. and Duke, W. B. (1977) An assessment of allelopathic potential in *Avena* germ plasm. *Weed Sci.*, **25**, 224–8.
17. Kowalski, R. (1982) Personal communication.

Chapter 5

Rotations and Crops

5.1 Rotations

THE rotation to be adopted on a unit should be designed by the farmer himself. He, after all, is the only person who knows the peculiarities of his land, the climate and topography, all of which have to be considered when a practical balanced system is developed.

The buildings and capital available for livestock and storage of crops also have to be taken into account at an early stage of the planning.

The crops grown must have a balance between those for human consumption – and therefore a more rapid source of cash income – and those required for livestock consumption. These latter crops will generally, but not invariably, be those which allow cleaning land of weeds or resting soil and restoration of soil organic matter. It cannot be stressed too much that the most effective way of obtaining weed and pest control is to alter the ecosystem of a field so that plants adapted to that particular ecosystem will not thrive under different management conditions. For instance, as discussed in Chapter 4, biennial and annual weeds which thrive in arable ground are eliminated in a grass field. Couch grass (*Agropyron* species), although rhizomatous and therefore well adapted to both an arable and grass plant association, will disappear from grass swards which are grazed, as the plant is incapable of producing leaf quickly enough to make food before further grazing.

To exploit the weaknesses of various undesirable plants which develop in grassland it is desirable to keep various classes of livestock as their different grazing methods affect plants in different ways. Under ordinary farm conditions it is usual to keep sheep and cattle which complement each other ideally. Incidentally, where it is feasible, hens kept on grass in movable folds provide not only a rich source of plant nutrient but also, by their scratching, aerate the sward. Horses are notoriously bad grazers and where these animals are kept in large numbers as in stables and studs the grass fields have to be topped either by sheep or by gang mowing.

The crops to be grown must be planned so as to allow time for adequate cultivations between each crop and to ensure that the succeeding crop will take advantage of the manurial residue from the previous crop – red clover would be followed by winter wheat, not spring wheat, as by the time spring wheat was

PLATE 6. *Animals are an integral part of holistic farming. Above*: Sheep grazing field of roots (C. B. Wookey). *Below*: Hereford cows on red clover ley (I. Fiedler).

sown all the nitrogen fixed by the red clover rhizobia would have been leached from the soil.

Sometimes it is advantageous to split part of one year's rotation between two crops so that in fact some of the land will only be used by a particular crop every other rotation. An example would be the growing of field beans in half the area of the root break.

The potential damage caused by insect pests, viruses and fungi must also be considered. Again the changing of the ecosystem will go a long way to reducing the commercial damage which they can bring about. The changing of system is, however, only really effective when the change is complete; the leaving of

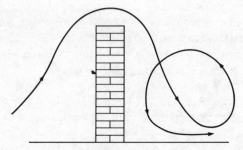

FIG. 5.1A   Brick walls cause severe turbulence on leeward side and wind erosion will still occur.

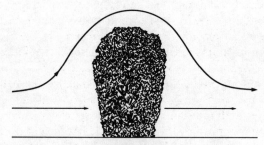

FIG. 5.1B   Hedges where about 30% of the wind can pass through at about ground level prevent turbulence and subsequent soil erosion.

self-sown cereals in the hedgerows, or diseased or hand-rogued potatoes in the headlands are both methods by which the pests can survive until such time as that crop returns in its rotation. Obviously the longer the rotation the less likely the carry-over, but for any given size of farm the longer the rotation the less the area of land in one crop, and this can give rise to difficulty in working the land. Again a balance has to be struck. Hedges are of vital importance in rotational farming; carrot fly can rarely rise more than 3 feet (1 m) above the ground and so it is physically impossible for the fly to infest new crops of carrots from existing crops as long as there is a normal hedge barrier. Besides the physical barrier provided by hedges, they also provide sanctuary for wildlife which is important in insect pest control. Indeed the ideal rotation is one where the whole needs of the farm (soil, plant, animal and man) are, as far as possible, met in a sustainable manner.

Rotations are very specific to a particular farm, and endless examples are of no real value. Consideration of two particular rotations, however, may be of value to the student as they show adaptation to topography and climate making full use of livestock needs as well as a cash crop income.

Rotation for Scottish intensively farmed land with bought-in store cattle for fattening
Year 1: Winter wheat. Cash crop following potatoes.

Year 2: Roots. For stock feeding – a cleaning crop after the cereal. Also a crop to receive some farmyard manure.

Year 3: Spring barley. Cash crop utilizing residues after removal of the stock fed root. Nitrogen level will be low so a malting sample of barley is possible.

Year 4: Grass seed. Hay for stock.

Year 5: Spring oats. For stock feeding.

Year 6: Seed potatoes. A cash crop. Plenty of time to get good deep cultivations before planting, and the application of large quantities of farmyard manure made in the previous year.

The rotation has half of the crops for stock and the other half for cash. The aim is for high-value cash crops – malting barley and seed potatoes. Potatoes will probably be able to be sold for seed as the wind speeds are too high to allow aphid damage so the possibility of virus infection is negligible. Livestock are purchased nowadays from the nearby stock rearers of the Highlands, but traditionally from Ireland (showing the multinational implications of holistic farming). The cattle are "finished" in covered cattle yards so using the barley, oat and wheat straw for production of farmyard manure. The oat straw would be fed along with hay to ensure the animals had adequate roughage intake. The oat grain, together with roots and reject potatoes, would complete the cattle diet. The nutrients supplied by this diet would be sufficient to ensure that the livestock were sold off "finished" in that one winter. In this type of farming rotation, each crop lasts only 1 year.

Rotation for a Cotswold farm with milk cows and sheep

Year 1: Spring wheat. A cash crop. Spring wheat is chosen to allow extra autumn grazing on the dry Cotswold land.

Year 2: Spring barley. After wheat the nutrient status is low and malting samples should be obtainable.

Year 3: Roots. Either kale or potatoes. The kale would be autumn grazed by the dairy cattle, so extending the grazing season. Potatoes would be "ware" for cash. Composted farmyard manure would be applied in early spring.

Year 4: Winter barley. This crop could be for stock feed or sale. It ripens early, so extending the harvest period. This crop could be criticized as a possible carry-over for fungus.

Year 5: Dredge corn or mashlum. Mixture of different cereals or cereals and legumes. The mixture reduces the likelihood of severe fungus or insect damage. The crop is used for stock feed. It would have been undersown with a 3-year grass ley mixture. These maiden seeds would give grazing for sheep after harvest.

Years 6, 7 and 8: Grass. For hay, silage or grazing.

This is a good rotation with roughly half of the crop for cash and the rest for livestock feed. If all the fields do not have a plentiful supply of water for the dairy herd, then those fields would be sown to hay/silage grass mixtures, whilst where there is water grazing strains would be sown. In the non-grazed fields fertility build-up, especially of potassium, would not be good enough, and so either liquid manure or composted farmyard manure would be applied.

5.2 Crops

It is not the intention to list every possible crop giving detailed instruction as to how it should be grown, nor to give a list of varieties, but to give information on the place of particular classes of crops in holistic agriculture and to indicate methods of meeting nutritional needs in an appropriate sustainable manner.

PLATE 7.   *Crops grown without the use of any agrochemicals. Above*: Winter wheat grown on chalk downland, yielded over 4.5 tonnes/ha (C. B. Wookey). *Below*: Oats grown in lucerne in low rainfall area of Australia, yielded over 3.5 tonnes/ha (I. Fiedler).

Cereals – The White Straw Crops

Wheat

Wheat is the most important crop in the temperate zones and it thrives where there is plenty of sunshine and about 60 cm of rainfall a year. The soils best suited are clay or heavy loams; light sandy soils are not suitable. Winter wheat has a special place in the rotation due to its need for autumn sowing, and as it has a high nutrient demand it is grown either after a heavily manured root crop or, more probably, after a 3-year ley or where permanent grass is being ploughed out. In these cases care should be taken not to have left the ground bare during August, or wheat bulb fly can be a severe pest. The fly lays its eggs on bare land in this period of the year and the newly hatched larvae are ready to damage the newly germinated wheat. It is usual to graze the ley hard during its last year to encourage clover and so get a high nitrogen fixation. Very recent developments indicate that it could be possible to direct-drill the wheat into a leguminous crop.

Winter feeding of the crop is not usually needed as the grazing will have raised the nutrient status, but occasionally, and especially after a long wet winter when the fixed nitrogen is likely to have been leached, the crop should be given a top dressing of aerated urine or slurry at the rate of 20 kg of nitrogen per hectare. If the crop is very "proud" after a mild winter then it can be lightly grazed, which will help tillering. Harvesting with a combine takes place when the crop is full ripe, and care must be taken to ensure that the machinery is correctly set. The loss of one grain per 1000 cm^2 is a loss of about 5 kg per hectare.

Barley

Barley does well on a poor soil, but the lime status must be adequate (a pH of over 6). If the nitrogen content of the grain can be kept low then there is a distinct possibility that the grain will be accepted for malting. If grain nitrogen level is low it is well worth while "dressing" the crop through a seed dresser in order to obtain a good even sample. The premium paid for malting samples adds greatly to cash returns, and the seed rejected by the cleaner can always be fed to livestock. Barley is often grown as a second cereal crop, the lower nutrient status of the soil being advantageous.

In appropriate agriculture, spring varieties are more frequently grown than winter types. The winter varieties need to be sown early and there may not be sufficient time after a root crop to prepare a seed bed. Additionally winter barley could be a host to over-winter fungus. The only occasion when there is sufficient time to prepare a good seed bed for winter barley is after grassland or legumes. However, the high nutrient status left by these crops would be better used by winter wheat.

On the lighter soils where barley is more frequently grown potassium can be a limiting macro-nutrient. The application of aerated slurry with its high potassium content is not possible because the nitrogen in the slurry would ruin any chance of obtaining a malting sample (and indeed could lead to very severe lodging). Traditionally the potassium level of the light soils was obtained by letting sheep graze the previous crop. Much of the nitrogen in the faeces and urine was leached before the spring barley was sown, but the potassium in the urine became attached to the clay domains in the soil, whilst phosphorus from the faeces was fixed and became available to the crop through symbiotic mycorrhizal activity. If this is not practical then potassium and phosphorus levels could be raised by the application of rock salts containing these nutrients – on alkaline soils the types of rock phosphates with aluminium salts are preferred.

Oats

This is the cereal for the wetter, poorer soils. It is the most resistant of all cereals to pests and fungi, but it does have a susceptibility to frost which is not usually a hazard in the wet western parts of the U.K. As it has a low nutrient demand it can be placed anywhere in the rotation, the richer nutrient status soils being reserved for wheat and then barley. Due to its allelopathic potential it makes the ideal nurse crop for clover and grass seeds. It is often grown as the last crop (when fertility is low) before the grass ley. The one-sided panicle, short-strawed varieties are the best to use as nurse crops as they offer little competition to the seedling clover and grasses. Oats are being used extensively together with the Wallace Soil Reconditioning Unit seeder as a method of growing a cereal for human consumption under arid or unstable soil conditions and even, in some cases, as a method of improving grazing in these very poor grasslands. In the U.K. where the crop is being grown for seed, and if it has not been retarded by frost, then it can be lightly grazed until mid-spring – say early April.

There is a strong demand for oats for human foods – muesli, rolled porridge oats and Sussex ground oats – and good grain samples will command a premium over ordinary grain. In the production of Sussex ground oats or pinhead oatmeal the husk is removed and sold as oatfeed. Oatfeed has a low feed value but it is of great use in lowering the excessively high energy of swill and other by-products used in pig feeding.

The botanical species *Avena nuda* does not have a high fibre content and shows promise as a high energy cereal source in the poorer climates. The oil in oats has a very high level of the essential fatty acid, linoleic acid, and can offer a useful contribution of the fatty acid in some animal diets.

Rye

By far the hardiest of all the cereals, rye stands very severe winters and summer drought, and thrives on poor acid soils. It is highly resistant to eelworm and disliked by rabbits and poultry. Because the crop is so hardy it is invaluable as a pioneer when rough grazings are being improved. The grain is highly prized for making into rye flour and rye bread, but the crop can be attacked by the fungus ergot and before feeding the grain to humans or animals steps must be taken to ensure that the grain is free of the black sclerotia.

The seed should be sown in the autumn on poor land and the earlier it is sown the better; it will then tiller and choke out annual weeds, and if the sowing is early enough the crop will have sufficient growth to allow good grazing and also high grain yield. The straw from rye is very hard, and so makes excellent bedding. Traditionally it was used for packing bananas and wine bottles, but today finds a ready trade for thatching and a use in horse racing stables as horses will not eat it, and so better diet control is obtained.

Dredge Corn

Dredge corn is a mixture of differing cereals grown together at one time to provide feed grain, the commonest being barley and oats, but any mixture can be sown. Because two or more species are sown then the actual plant density of any one species is reduced and diseases specific to one will not spread as easily as if only one cereal type had been grown. It is general to sow at a normal cereal rate – say 175 kg to the hectare – but the actual proportion of one cereal to another depends on local conditions, pests, climate, etc. As it will not be possible to get the cereals ripening at the same time yield will be lowered, but not by as much as if disease had struck a crop. However, it is usual, because in general oats ripen before barley, to sow late-ripening varieties of the former and early-ripening varieties of the latter. The mixtures are almost always spring sown.

Legumes – The Black Straw Crops (Beans, Peas, Lupins, Mashlum)

Beans

Beans may be either winter or spring sown. Which type to grow will depend on climatic conditions and land availability. Winter beans will outyield the spring varieties but the spring varieties are less susceptible to the fungus, chocolate spot, and will indeed yield more crude protein per hectare than the winter varieties. They thrive in mild climates and do best on the heavier soils which have a high pH. Beans will not grow well on light acid soils. They are usually grown as a cleaning crop, as the space left between rows allows of inter-row cultivations. Where livestock are likely to need dry proteinaceous foods they can replace half of the root break, and as they are legumes they have the ability to fix nitrogen. As they are a good cleaning crop, and because of the residual nitrogen from fixation, they are a good preparatory crop to winter wheat. Like all legumes they have a high requirement for phosphorus, and as they also have a need for a high soil pH basic slag, where available, is a good seed bed fertilizer. In the absence of basic slag well-composted farmyard manure should be added to the seed bed.

Winter beans should be sown in late autumn (October), and spring varieties should be sown as early in the year as possible (February). This early sowing gives good establishment and they are not likely to suffer black fly attack which occurs in May and June. The fungus diseases, chocolate spot and stem rot, can under some conditions be troublesome, but if the crop is widely spaced and not too thickly sown the consequent circulation of air through the plants will lessen the likelihood of the diseases assuming economic importance.

Beans can be a difficult crop to harvest as they do not ripen uniformly. Some farmers cut the crop with a windrow mower and then combine from the windrow. In some seasons there can be severe shedding of seed, and if this happens the stubbles should be folded by sheep or pigs. In some situations it may not be possible to fold off the shed seed, and under these conditions the seed should be allowed to germinate and be treated as a green manure.

Peas

Peas prefer a mild dry climate, with a medium to low rainfall. They do best on light soils which must be well supplied with lime. Like beans, they have a high requirement for phosphorus and if there is a phosphorus deficiency in the soil then basic slag or rock phosphates should be applied as the seed bed is being prepared. Excessive rainfall can give too much haulm growth, which can lead to problems at harvest which, at the best, is never easy. For the freezer trade specialist machines are used, but for ordinary farm use it is best to cut with a windrow mower and follow with a combine harvester.

Lupins

A crop which historically was grown in the eastern counties of the U.K., lupins appear set for a revival in Europe, and indeed are potentially the most valuable legume for the holistic farmer. The original, pre-1939 war, varieties were "bitter" and contained alkaloids which were poisonous to both livestock and humans. Intensive plant breeding, especially in Australia, has developed "sweet" alkaloid-free varieties. They are the only legume crop which will tolerate acid soil conditions and indeed do best on thin poor sandy soils. They therefore make, together with rye, an ideal pioneer crop.

The grain of the new varieties can contain almost as much protein as soya beans, but unlike soya beans will grow well in temperate conditions. Unfortunately the lupin protein is low in the amino acid, methionine; therefore if the lupin is to be used in poultry diets it must be used in conjunction with methionine-rich proteins. The relatively high oil content of the seed gives it a high energy feeding value although the seed husks can cause problems in young pig diets. They are an ideal food for all ruminants, the protein being relatively undegradable and the high fibre content helping to stimulate good rumen function.

The newest Australian varieties[1] can withstand up to 7 °C of frost, and indeed many varieties do have a vernalization requirement; nevertheless they should not be sown until there is no possibility of frosts of this severity occurring. However, maximum grain yield is only obtained with a long growing season and in the favoured warmer parts of the southwest of the U.K. the crop can with advantage be autumn sown.

The crop utilizes phosphorus to great advantage and the place in the rotation may be dictated by a high soil phosphorus level. The crop will not, in the early stages, tolerate competition from weeds and so cultivations to give a weed strike are appropriate. Lupins produce a great deal of green top growth and harvesting should be by mowing to windrow and combining from the windrow. The green matter can be fed to sheep as long as no heavy dews occur during the feeding period: dew appears to allow the formation of lethal alkaloids in the green material. The lupin stores a very high level of phosphorus and potassium in its deep tap roots and the crop is therefore a way of concentrating these nutrients.

Before sowing, it is advantageous to dress the seed with lupin-specific rhizobial bacteria, and if this is done the quantity of nitrogen fixed by the crop is more than adequate for the grossest nitrogen feeders.

Mashlum

Mashlum is the name given to a mixture of cereal grains and legumes which can be used either as a source of protein-enriched grain for livestock feeding or may be cut at immature stages for silage. The crop exhibits all the benefits of a dredge mixture, and moreover as plants of an entirely different botanical species are incorporated into the mixture it is unlikely that disease damage will occur. It is important to observe several basic principles if mashlum is to be grown. The cereals usually consist of barley and oats but in some instances, where the harvest grain is for poultry feed, spring wheat may be included. Beans would be the legume of first choice as they are erect in growth and unlikely to cause lodging of the crop. Peas may be used in the mixture but at a low level. The soil lime status must be correct and, as with legumes and cereals, the phosphate and potassium levels should be at a satisfactory level. The types of mixture used for different needs are as follows (quantities per hectare):

1. For silage:
   100 kg spring oats
   75 kg early-ripening spring barley
   75 kg peas
   (This mixture would be cut when the grain is at the milky stage.)

2. For grain for stock:
     100 kg oats
      25 kg barley
      75 kg beans
      25 kg peas
3. For poultry feed grains:
      50 kg spring wheat
      75 kg spring oats
      50 kg spring barley
      50 kg maple peas

In the cooler climates it would be usual to sow the beans in any mixture about 3 weeks before the cereal part of the mixture; the time differential being necessary to ensure that both bean and cereal ripen at about the same time.

Roots for Cash Crops – Potatoes and Sugar Beet

Potatoes

Potatoes will grow well in all temperate climates but are at their best in areas with lower temperatures and regular rainfall. They should not be grown where late frosts could give a total crop loss. Soil should ideally be deep, easily worked, well drained and slightly acid. The best crops cannot be grown on heavy clays. In the rotation they are taken in the root break, and because of inter-row cultivation and the crop's ability to smother weeds, they are an excellent cleaning crop. Heavy application of composted farmyard manure – as much as 25–30 tonnes per hectare – ensures an adequate addition of all macro- and micro-nutrients and the organic content gives a better tilth, holds water and yet keeps the soil open. The dark colour of the organic matter will help to warm the land. The potato is a gross feeder and the application of aerated slurry after planting, but before leaf emergence, will give additional quick-acting nitrogen for yield as well as potassium for improved leaf efficiency and disease resistance. If the slurry is applied after the leaf has emerged, scorch can occur, although this is minimal if the slurry has been well aerated before application. The application of 75–100 kg of nitrogen per hectare would not be considered excessive. If potatoes are grown in areas of high windspeed there is very little likelihood of aphis-carried virus occurring, and the crop could possibly be sold for "seed" rather than "ware". The fungus, blight, as every schoolchild knows, caused the Irish potato famine, and indeed it can in certain years be a considerable problem if it occurs early. Under these conditions or in areas where blight is always expected early then the use of Bordeaux or Burgundy mixture as a preventative is justified. The spray should be applied to the undersides of the leaves (where the fungus enters the plant through the stoma) at the end of June. Late attack of blight is dealt with by mowing off the haulms, removing them from the field, and only if composting methods are good should they be used for compost. The crop should be lifted as soon as is convenient.

In some years there is a potato glut, and it is then worth considering turning the surplus small tubers into potato silage or, more advantageously, ensile about 1 tonne of potatoes with 3 tonnes of grass. Potato haulms are toxic to livestock and should not be ensiled.

Sugar Beet

The growing of sugar beet is often controlled by the local sugar beet authority, who may insist on chemical spraying and other unacceptable non-sustainable methods. If this is the case then the crop is not available to the holistic farmer. (Indeed many holistic farmers would ponder the wisdom of growing crops which are better suited to third world countries, who are in desperate need of export earnings.) However, if there are no unacceptable conditions attached to cultivations (and in some countries the holistic farmer may decide to grow the crop) then not only is it a good cleaning crop with a cash return, but also the farmer may buy back from the processors sugar beet pulp, an excellent animal feed. The crop is not therefore a drain on nutrient reserves.

The crop needs early moisture to establish the seedlings, and then long periods of sunlight and good rainfall, the former to allow of full photosynthesis and the latter to maintain good growth. The best soils are deep, easily worked loams with a pH above 6. Heavy clays are not suitable as removing the crop in wet autumns can destroy soil structure. It is a good cleaning crop and takes its place in the root break part of the rotation. It responds to heavy applications of well-composted farmyard manure. Manure which is not composted, or which is fresh, will encourage the development of fanged roots which lower the sale value. As with potatoes, application of aerated slurry to the growing crop will enhance quality of the leaf activity and therefore give better sugar production, whilst the nitrogen present will increase yield. As the basis of payment for sugar beet is on a combination of yield and sugar content then this application is very cost-effective.

Harvesting is carried out according to the demands of the local sugar beet factory, but best financial return is made when the crop is lifted at full maturity; in the U.K. at the end of October.

The beet tops should be taken from the field and allowed to wilt, so reducing the oxalic acid content which can be toxic, before feeding to cattle and sheep.

Roots for Livestock – Fodder Beet and Mangolds, Swedes and Turnips

Fodder Beet and Mangolds

These crops are of the same botanical species as sugar beet and will be grown under similar conditions. Fodder beet, with its high dry-matter content, is replacing mangolds as a crop. As a cleaning crop they are grown as part of the root break. As long as the lime status of the soil is satisfactory, they respond to heavy applications of compost and aerated slurry. Fodder beet is very palatable and finds a place in the diets of ruminants and pigs. Traditionally it was developed as a pig feed.

Swedes and Turnips

It was the introduction of the turnip into the U.K. in the mid-seventeenth century that revolutionized agriculture. It became possible to over-winter stock and therefore the pioneer cattle and sheep breeders were able to make rapid strides in livestock improvement.

These root crops may be grown for folding off by sheep in the field where they were grown, or more frequently are lifted and fed to animals being over-wintered in the stock yards. Recently hybrids have been developed which can be sown into corn stubble to give either a green manure or better stubble grazing.

Swedes and turnips do not differ significantly in their growing requirements but in general swedes are more frost-hardy and require a longer growing season (and are therefore not used for green manures). They have a higher dry-matter content which, together with their higher yields, give considerably more dry matter per hectare than turnips. Finally, swedes tend to have a higher sugar content and are therefore more palatable.

Both crops are susceptible to mildew, and whole plantings can be destroyed by the flea beetle. To avoid mildew, sowing after the end of April (late spring) appears most satisfactory. If the seed bed has been well fed with compost, germination will be rapid and the plant will be well established before flea beetle attacks; strong-growing plants are more able to withstand the ravages of the flea beetle. If, however, the attack is severe the use of derris may be justified. Machines are now available to take away the very heavy labour of harvesting the big tonnages of these crops which will be obtained under proper husbandry methods.

Stubble turnips are broadcast into stubble as soon as possible after removal of grain and straw, and in a good season useful keep will be available for livestock. In particular, because of nutrients obtained from the shed grain and the stubble turnips, the grazing of the crop by ewes has a flushing effect with its consequent improvement in the rate of conception. Where black straw crops have been grown the sowing of stubble turnips will ensure that the nitrogen fixed by the legumes is not lost by leaching.

Forage Crops for Livestock – Kale, Rape and Cabbage

Kale

The three main varieties of kale, marrow stem, thousand-headed and hungry gap, if fed in succession can provide green keep for ruminants throughout the whole of the winter months. Marrow stem, which is the least hardy, can be followed by thousand-headed, which will last until the end of late winter (February), whilst hungry gap can be fed from the end of the thousand-headed well beyond the arrival of the new spring grass.

All varieties are gross feeders and are usually grown in the root break where, if well supplied with nutrient and kept free of weed in the establishment period, they will smother out any further weed development. However, as they are all resistant to wireworm attack they can be grown where old pastures, known to be infected by the pest, are to be ploughed up. They develop a good tap root which will break up old turf very effectively. The best soils are the medium loams which receive a high rainfall, but the crops will grow on any type of soil, even with low rainfall, as long as the lime status is satisfactory. If the crop is grown on heavy soils it cannot of course be folded off in winter as livestock will have been yarded to avoid damage to soil structure. Under these conditions the crop has to be carted each day, and anyone who has cut kale on a cold wet or frosty day would not wish the job onto his worst enemy! Where an electric fence is used to fold stock over the kale it is very worth while planting two rows of cabbages where the electric fence will eventually be placed, so avoiding the possibility of electric short-circuit by the kale leaves.

The crops must be fed well, and will respond to the maximum quantity of compost or well-aerated slurry (or both) that can be spared. It appears quite impossible to overfeed the crop, although very occasionally under very adverse weather conditions the crop can lodge.

Marrow stem kale (and for that matter cabbage) stumps make future cultivation difficult: they can easily be removed by "spinning" them out with an old-fashioned type of potato spinner.

Rape

Rape is very similar to kale but gives very heavy yields in cool moist conditions. The cultivation methods would be the same as for the kales. Rape is often used as a catch crop; it can be grazed 2 months after sowing, and used this way is, like stubble turnips, a superb crop for the autumn flushing of ewes. It is often used as a nurse crop for grass seed mixtures where direct reseeding is being practised. The early growth it gives provides keep, and the actual grazing of the newly germinated grass seeds mixture gives good consolidation of the ley: further, the grazing will remove any show of weeds.

Good winter grazing can be obtained by sowing in the autumn 2.5 kg rape, 5 kg Italian rye grass and 7.5 kg trefoil per hectare.

(All readers will be aware of the increased acreages of rape being grown for vegetable oil production. In conventional agriculture it is mainly used as a break crop – as has been argued, a concept unknown to the holistic farmer – from the monoculture of cereals. The winter-hardy varieties, which give the greatest yields, have to be sown in the very early autumn (July) so as to be well established before the onset of adverse winter conditions. Its place in a rotation would be instead of winter wheat, but because of its early sowing the preceding grass would have to be ripped up earlier than for wheat, so depriving the farmer of grazing in what may be a dry time. The crop, when grown, is very susceptible to attack by many predators which are presently only controllable by spraying – an avenue not open in sustainable agriculture. It is not therefore a recommended crop for the holistic farmer.)

Cabbage

Cabbage can be grown as food for human consumption or livestock and will, under good conditions, give the highest yield of any forage crop. Well-drained soils supplied with lime, ample plant nutrient and a climate which gives plenty of gentle rain and long hours of sunshine can give yields in excess of 100 tonnes per hectare. The cultivar grown will depend on the time of year that the mature plant is required for farm or market needs. The response of the crop to well-composted manure and well-aerated slurry seems limitless.

Flea beetle can be a severe problem, so it is worth sowing seeds in nursery beds where inspection is easy and derris can be easily and safely applied should the need arise. The young plants are transplanted into growing sites in the root break, although some Dutch white varieties have been successfully transplanted into grassland. Transplanting of cabbage plants seems to be one of the few farm jobs which can be carried out better by machine than by hand. The distance between plants and between rows will depend on the variety, but it is well worth giving plants plenty of space to explore the available soil for nutrient.

Inter-row cultivation by steerage hoe will keep weeds to a minimum, so allowing the crop to perform its cleaning function in the root break.

The folding of cabbage by livestock can be controlled by electric fencing, and as the crop is low-growing there is little risk of short circuit due to leaves touching the wires.

Trees

The farming of trees is not a new idea: as long ago as A.D. 1198 Royal Decrees granted to villagers rights to gather willow for firewood, and at the same time to use the common land for grazing. The Keyline principle, and its suggested use of shelter belts consisting of native species, opens up the real possibility of using trees as a crop in holistic agriculture.

Coppicing

Most broad-leaved trees can be coppiced. Research work carried out in Europe and Canada shows the great value to be obtained from growing willow and poplar (*Salix* species) in cooler climates.

PLATE 8. *Coppicing*. *Above*: Typical area recently coppiced. Self-sown oak and ash for larger timber (Warden, Bradfield Woods). *Below*: Willow stripped to help drying. Ash poles for handles of pitchforks and scythes. Technical foliage on left of picture could be used for muka (Warden Bradfield Woods).

Willow: In the U.K. willow has, under a coppice system, yielded over 37 tonnes of fresh-weight wood per hectare per annum over a 4-year coppice period.[2]

Coppicing can take place over varying numbers of years, the actual length of time between each cutting depending on the use for which the crop is to be used. The trees regenerate very rapidly and growths of up to 2.5 cm per day over a growing season are frequently recorded. The life of a willow plantation is at least 50 years and the trees do not appear to impoverish the land. After removal of the root stumps the fields can be returned to arable crops. The trees appear to grow on any soil type, thriving in the worn-out peatlands of Ireland as well as the heathlands of northern Scandinavia. They give the best response on good deep working loams, such as those found in the Somerset plain where the plant has been grown for generations as a source of basket osier.

Where energy is the main object the coppice is usually cut every 4–5 years, and the wood is usually burnt directly in stoves. However, the wood can be subjected to destructive distillation in retorts and under these conditions gas, which can be used as fuel, and chemical liquors are produced. The residue, a type of charcoal, can be burnt as a heat source. The liquors produced by the process are raw material for the chemical industry, producing many of the chemicals at present produced from oil.

Annual coppicing gives the "rods" used by basket makers. At the other extreme, if coppicing is restricted to a 10-year period the wood can be used in the paper pulp industry producing a kraft similar to that obtained from poplars.

The willow has a very wide genetic variability and it is important, if the best results are to be obtained, to use the correct species or hybrid. Local forestry stations should be able to advise on the best variety for a particular purpose in any particular climate. The common wild osier of the U.K., *Salix viminalis*, is extremely vigorous and frost-hardy and appears to be one of the best varieties for British conditions. Plantations are very easy to establish from cuttings. The cuttings, about 20–30 cm long, are planted into grassland; the distance between rows varies, but to allow mowing between the developing plants it is advisable to leave at least 75 cm. Distance between plants in the row again varies according to crop use, the smallest distance being for osier production where the distance would be as little as 30 cm. The cuttings taken from 1-year-old shoots, called rods, are planted in March or April to a depth of about 15 cm. In the first year the grass between plants should be kept mown so as to reduce competition with the young cuttings. First- and second-year growth has to be removed to help the "stool" to form, and it is from this stool that future commercial growth will develop. In the third year the stool is well developed and coppicing can be considered.

Sweet chestnut: This is an ideal coppice plant, usually planted at 2–3 m intervals. The young growth is much in demand as fencing posts and firewood. The timbers are frequently split into pales, wedge-shaped, and when these are bound together with wire the flexible fencing, chestnut paling, has a great number of uses as stock- (and human)-proof temporary fencing. Large timbers tend to split and it is not therefore a suitable furniture or building timber. The bark is very rich in tannin and in some countries it is used in the leather industry, and in warmer climates is grown for its fruit.

Eucalyptus: There are at least 500 different species of eucalyptus and there is sure to be at least one species for any particular condition, from the arid hot climates to the temperate zones. Some varieties produce superb timber; some are more suited for paper pulp and others can be coppiced for firewood. In the British Isles the variety *E. gunnii* will survive long periods of low temperatures ($-10\,°C$ to $-14\,°C$) and will even survive short periods of $-16\,°C$. *E. nitens*, an extremely good firewood-producer with rapid growth, is not so hardy as *E. gunnii* but will survive long cold spells down to $-6\,°C$ with short periods not exceeding $-9\,°C$.[3]

Leaves borne on young wood are round, and as long as the tree is pruned this is the only type of leaf produced, and is much in demand by the florist trade. If the tree is allowed to grow unattended the leaves adopt the typical long spear shape, and the plant will eventually flower and produce high yields of very good feed for bees.

Trees for Quality Furniture Timber

Oak: The greatest of the British trees is still planted commercially, although it will take at least 100 years to reach maturity. In the past the tree was frequently pollarded when the trunk was about 2.5 m tall. The new growth which developed from the crown of the pollard was then used for fencing posts or firewood. The purpose of pollarding at that particular height ensured that the browsing animals did not damage the new growth. Trees for the best timber are of course allowed to grow to their full height with only trimming of surplus branches to keep the tree growing in an erect habit. Oaks thrive on the heavy deep clay soils where they are assured of a plentiful supply of water.

Beech: The beech does not naturally spread into the colder climates as considerable warmth is required to ripen seed. It has a very shallow but extensive root system which is well adapted for growth on the thin soils of the chalk and limestone hills of southern England. Like the oak the wood is favoured for furniture, especially chair manufacture, but again like oak was often grown in the past as pollards for the supply of firewood.

Ash: Another native British species, the ash grows on the better land and is prized for use as handles for hammers and other hand implements. Nowadays most of the timber is used for games equipment such as hockey sticks and oars. It has a ready demand for quality furniture, whilst the tops and offcuts make excellent firewood. The timber quickly deteriorates under the temperate weather conditions and so does not make satisfactory fencing posts.

Trees for Fruit

There is no reason why standard, half-standard or even bush orchard fruit trees should not be grown on the Keyline principle. For the cultivation of these trees the reader is referred to any of the good textbooks on this subject.

Walnut: The timber of the walnut is in great demand for the furniture trade but, due to scarcity and expense, fine trees are usually turned into veneer. The fruit is in great demand for the nut trade: under British conditions, because of the grey squirrel and cool summers, most fruit is picked in the green stage and used for the luxury pickled walnut trade.

Mulberry: The mulberry is not a native European tree, but has been introduced quite extensively, varieties from both Asia and America doing well under British conditions. The trees are either male or female. It is usual only to plant female trees as they are capable of producing the delicious purple fruit without fertilization, although of course the fruit does not produce seed. Traditionally the mulberry was planted as a specimen tree in parkland or large gardens, but there is no reason why it should not be used under Keyline systems.

Leguminous Trees

In the U.K. we do not have any native leguminous trees. However at least three varieties have been introduced and will grow well under British conditions.

False acacia (Robinia sp.): This is a decorative tree, bearing racemes of white flowers in the spring. It is almost entirely confined to gardens but, planted in Keyline situations, the benefit of the nitrogen fixation would be of use to other trees in the plantation.

Honey locust (Gleditsia): The honey locust is almost a weed in some parts of the world, but it has the advantage, being a legume, of fixing nitrogen. Under warm conditions the long plump pods will ripen, filling with sweet pulp which is readily consumed by all kinds of animals, including man.

Laburnum: The golden racemes of this plant are exquisite, but the tree has no place in agriculture because of its poisonous properties.

Other Trees

Poplars: These rapid-growing species, members of the same family as the willow, can be invaluable where windbreaks are urgently required. The trees establish quickly from cuttings, which can be as long as 2 m. If the trunks are kept free of side shoots the timber can be shaved to produce material to make tomato chips or punnets for soft fruit. The timber is still unsurpassed for use in matchsticks and matchboxes. Timber of poorer quality is used extensively in the paper pulp industry. The trees should be planted at 4–5 m on the square.

Lime (Linden): These are magnificent trees which will enhance any forest or copse. The timber is of golden colour and was, in days past used for the finest carvings (it appears to be almost the only wood used by Grinling Gibbons). However, nowadays the greatest value is as a source of nectar for bees, and in Europe the flowers are dried to make the famous French tillia infusions for hot drinks.

Conifers: There are several native British conifers, and in conditions where the broad-leaved trees so far discussed will not thrive any of these will make good alternatives, provided they are planted so that they can live happily with other members of the ecosystem. Conifers will also often thrive in association with broad-leaved trees, and if used sparingly on mixed plantations those which are evergreen give year-round windbreak protection. The preferred tree would be the Scots pine because of its splendid timber. The larch and spruce, to be harvested as pit props and paper pulp (and there is already an over-supply in the U.K.), must be close-planted; and who has not observed the dearth of other life in the Forestry Commission plantations of Scotland and Wales? As trees for cropping they therefore have little place in Keyline planting though their fast growth might make them attractive in some situations.

Besides the trees which have been mentioned, there are of course many other species (hazel comes immediately to mind for coppicing and fruit), all of which can have a place in agriculture. It will depend on topography which plants will be used but, as always in holistic agriculture, consideration will be given to the benefit any particular tree contributes to balance in the ecosystem.

Mycorrhiza and Trees

It has been known for some considerable time that the mycorrhiza, both endotrophic and ectotrophic, establish symbiotic relationships with both coniferous and broad-leaved trees. It is generally accepted that this particular symbiosis is extremely important in that it provides the tree with the majority of its phosphorus, potassium, and magnesium requirements. Tree management must be such the mycorrhiza flourish.

In parentheses recent work in the Netherlands[4] has shown that American oaks, Austrian and Corsican pines and larches are dying in areas which are in close proximity to highly intensive livestock units. Examination of the trees shows that the bark of the oak trees had become detached from the trunk, roots had not properly developed, were smooth with no root hairs, and little or no mycorrhizal infection was found. The same workers have monitored the presence of the mycorrhiza in woodlands and note that in two of the affected woodlands no mycorrhiza could be found in the soil. In another woodland also close to an intensive livestock unit in 1980 they identified 63 different species of the Basidiomycetes fungi and 24 different mycorrhiza, but only 3 years later in the same woodland they were only able to find 40 species of Basidiomycetes and 11 mycorrhiza.

The pH of slurries from intensive livestock units are alkaline (pH over 8) and it is possible that seepage of these slurries into woodlands is inhibiting or killing mycorrhiza (mycorrhiza do not thrive in alkaline soils), or that the ammonia gas given off by slurries may be falling as alkaline rain in these woodlands.

Muka

Wherever trees are grown there comes a time when they have to be felled for timber, or cut for coppice wood, or pruned. After the usable timber has been removed there are left the twigs and leaves and petioles or pine needles which

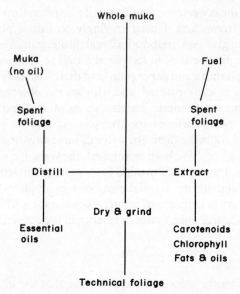

FIG. 5.2   Options for processing technical foliage. (After Barton[5].)

are generally burnt. However in the 1950s scientists in the Latvian state of the U.S.S.R. discovered that if pine needles and twigs were dried and ground they made excellent animal feed. Later, work was carried out using leaves, petioles and twigs of deciduous trees. The maximum diameter of twigs allowed into the mixture for drying is 0.6 cm. Above this diameter the twig has too thick a bark, and this reduces the feeding value of the final product. The mixture of leaves, petioles, twigs and pine needles before drying is called either technical or commercial foliage. Once dried the product has been named by the Russians muka. Muka is strictly only obtained from pine needles and coniferous twigs; other dried technical foliage is described by the species from which it is derived – for instance oak muka or birch muka.

Table 5.1 gives the compositional analysis on a dry matter basis for various types of muka as well as feed energy values calculated from the analyses. The leaves and twigs of several of tropical legumes are reported to have crude protein levels in excess of 20% and some temperate poplar hybrids average 19.5% crude protein on a dry matter basis.

Besides the nutrients for animals, muka has been shown to contain:

1. Essential oils: pine and eucalyptus essential oils are used in cosmetics and medicine.
2. Vitamins: vitamins A, C, E, K, provitamin D and riboflavin have all been identified.
3. Salicin: found in willow and poplar; is used as an analgesic and antipyretic.

4. Carotenoids: yellow and red pigments associated with chlorophyll.
5. Flavonoids and Tannins: these chemicals, when treated with formaldehyde, develop adhesive properties which could find a place in wood bonding.

TABLE 5.1

Type of muka	Percentage crude protein	Percentage acid detergent fibre	TDN	ME MJ/kg
White Spruce[5]	5.9	34.1	57.3	*
Poplar[6]	9.7	*	*	9.7
Heather[6]	12.9	*	*	6.0
Ash	10.3	30.1	53.0	9.3
Oak	10.6	33.0	48.0	8.5
Willow	10.1	37.0	48.0	8.5
Genista	16.0	29.0	57.0	10.0
E. gunnii	11.1	19.7	49.0	8.6

* Figures not available.

In this book the interest in muka is mainly as an animal feed. Published Russian experimental work in the field of animal husbandry is summarized in Table 5.2. These results clearly show that all classes of livestock can be fed diets containing 5% of muka with no deleterious effects, and indeed the feeding of this quantity in the diet appears to improve animal health. Experimental work in North America appears to confirm the non-deleterious effect of muka in diets. Table 5.3 shows results.

The finding concerning health status may at first seem surprising but, on consideration, this improvement may be due to a raising of the vitamin levels

TABLE 5.2   *Reported benefits from use of muka in the U.S.S.R.*

Class of livestock	Recommended supplement level (%)	Acceptance level (kg/ head/day)	Benefits
Poultry	5	–	Reduced susceptibility to disease, increased (18%) weight, increased egg production
Cattle	5	1–2	Reduced susceptibility to disease, increased vitality, increased (15–20%) weight; easier gestation
Milk cows	5	1–2	Healthier cows, increased (12%) milk production
Pigs	5	–	Increased (15%) weight
Sheep	–	0.25–0.50	Improved growth, increased
Goats	–	0.25–0.50	productivity, healthier animals,
Horses	–	1.00–2.00	improved reproductive capacity

Adapted from Keays, J. L. and Barton, G. M.[7]

TABLE 5.3 *Summary of muka feeding experiments in North America*

Test animal	Foliage	Form of foliage	Treatment of foliage prior to feeding	Details of feeding experiment	Results
Chickens	Loblolly pine	Needles and technical foliage	Dried and ground	Rations were formulated at 2.5 and 5% levels and compared against alfalfa controls	Did not adversely affect growth, feed conversion or mortality. Taste panel judged quality equal to controls[8]
Laying hens	Loblolly pine	Needles and technical foliage	Dried and ground	Ration was formulated at 20% level and compared with standard layers diet	Egg quality was as good as control diet. No adverse flavour. Lower production due to lower energy level of test diet[8]
Broilers	Red spruce, balsam fir	Needles	Dried and ground	Ration was formulated at 5% and compared with a standard corn high-energy diet	The growth rate was depressed, feed consumption was not affected. Taste was not affected[9]
Sheep (wether)	Red spruce, balsam fir	Technical foliage	Ground and heated to drive off volatiles	Ration was formulated at 25% level and compared with timothy hay	The addition of muka to timothy hay did not affect palatability. The mixture had a significantly lower digestibility than timothy alone[10]

Adapted from Goldstein I. S. 1981.

in the diets which, under Russian conditions, could be lower than that in other countries. Again it is known that tall oil – a by-product of the wood pulp industry – is rich in linoleic acid, and it is not unreasonable to postulate that the triglyceride fraction of muka is also high in linoleic acid; linoleic acid is an essential fatty acid (it also increases egg weight in poultry). It is established that some (eucalyptus) mukas have fungicidal properties. Do all mukas exhibit this property? Do these fungicides control fungi which may be on or in grain? Do some mukas have properties which inhibit the adverse effects of mycotoxins which we know contaminate many grains? Many questions have yet to be answered, but sufficient is known to ensure that this raw material has an exciting future in animal feeding. It will be important for each country to test various mukas for toxicity, palatability and product taint. Up to the present the only muka reported to have unfavourable properties is the ponderosa pine, which when eaten by cattle can cause abortion.

Yield of Muka

About $7\frac{1}{2}$% of the weight of a tree is in the form of dried muka. The yield of muka will of course vary according to whether the woodland is broad-leaf or coniferous, and the density of the original planting, but as a rough guide a coniferous plantation will, when being felled for timber, produce about 40 tonnes of muka per hectare. This would be sufficient for incorporation into 800 tonnes of animal feed. A traditional broad-leaf coppice could produce about 5 tonnes of muka per hectare annually. On the other hand willow or poplar grown specifically for muka production, on land unfit for other crops, and harvested each year could produce at least 12 tonnes of muka per hectare per annum.

References

1. Reeves, T. (1984) Director of Rutherglen Research Institute, Victoria, Australia (personal communication).
2. Robertson, A. (1984) Willow plantations in agroforestry. *Span*, 27(1), 32–4.
3. Anon, (1981) *A Key to Eucalypts in Britain and Ireland*. U.K. Forestry Commission Booklet No. 50.
4. Jansen, P. J. (1985) Zure regen en paddestoelen, een verkenning in de peel. *Coolia*, 28(1), 13–16.
5. Barton, G. M. (1981) *Organic Chemicals from the Biomass* (editor Goldstein, I. S.), Chapter 11, pp. 250–280. CRC Press, Boca Raton, Florida, U.S.A.
6. Anon, (1980) *Nutrient Allowances and Composition of Feeding-stuffs for Ruminants*. Booklet 2087, Ministry of Agriculture, Fisheries and Food.
7. Keays, J. L. and Barton, G. M. (1975) Can. For. Services Prod. Lab. Inf. rep. No. VP-X-137, and *Appl. Polym. Symp.* (1976) 28, 445.
8. Watts, A. B. and Manwiller, F. G. (1981) Evaluation of Southern Pine needles and twigs as a poultry feed supplement. *For. Prod. J.* (submitted).
9. Gerry, R. W. and Young, H. E. (1977) A preliminary study of the value of conifer muka in a broiler ration. *Res. Life Sci.*, 24(2), 1.
10. Agpar, W. P. *et al.* (1977) Estimated digestibility of conifer muka fed to sheep. *Res. Life Sci.*, 24(1), 1.

Chapter 6

Grassland Management

6.1 General Considerations

IN the British Isles if man were to stop cultivating his arable land it would revert first to grassland, then scrub, and finally establish as woodland, the natural ecosystem of the country. As grassland is in an intermediate ecosystem it is difficult to manage; under-stocking allows in the scrub; over-stocking turns it into bare land. It is only necessary to consider the bare ground which surrounds water troughs and pathways and gateways in grassland to realize how great is the removal of grass where heavy animal traffic occurs.

The output of herbage from a pasture will depend not only on the stocking rate and the general level of plant nutrients in the soil, but also on the air and soil temperature and the supply of water. Stocking rate and nutrient levels can be controlled but the other conditions are not amenable to regulation; therefore there is a natural tendency for flush growths in spring and early summer and in the autumn. In late summer output drops due to drought conditions, and in winter the cold does not allow growth in most parts of the U.K. Of course, in other parts of the world growth may be at a maximum when winter rains occur, and may cease under summer drought. Flushes of growth will cause a great deal of wastage of valuable animal feed.

Grassland management is a technique developed in an attempt to even out the production of herbage, and at the same time maintain the pasture in the desirable balance between woodland and arable associations. In some countries the natural ecosystem is grassland, and farmers in these parts do not have as difficult a job as those farming where the grassland state is intermediate.

In practice the maintenance of the intermediate grassland state is attempted by the use of grass mixtures which will give very early spring growth, or the use of grasses which will grow under summer drought conditions. The balance between woodland and arable associations is obtained by deciding whether in any particular year the grassland should be used for grazing or conservation (hay or silage making), and if for grazing the rate and type of stocking for the land.

Figure 6.1 clearly shows the alterations which can be made in sward composition by changing the stocking rate.

100

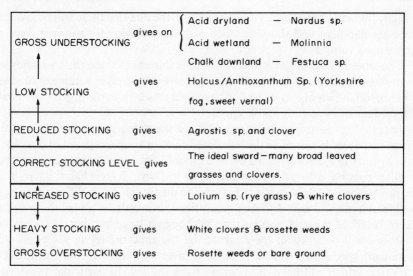

FIG. 6.1

In the British Isles the deep-rooting herbal ley is an integral part of the system on most ecological farms. Many of the conditions for making a good ley are common to both conventional and organic farms, but the importance of obtaining good results is such that it is worthwhile considering the whole subject rather than simply to confine attention to the difference between the reductionist and holistic approaches.

Apart from the ley's own value as a crop, the period of ley is a time when

1. soil organisms can work undisturbed and so build up fertility;
2. leguminous plants sown with the grass can build up a supply of nitrogen through the rhizobial symbiosis;
3. grasses and deep-rooting herbs can bring up from considerable depths minerals which are of value to the grazing herbivores;
4. roots of these deep-rooting plants, when they die, leave channels in the soil, and these channels assist drainage and aeration.

It is therefore of very great importance to get the best out of the ley, and every care must be given in order to get the grasses well established.

The length of time for which a ley is to be left down is decided by the general farming policy and rotation, but it should be at least 3 years: yield is lowest in the first year and increases to reach a peak in the third and fourth year, so the annual average yield will be higher for a 4-year ley than for a 3-year ley. Good seed mixtures are expensive and there is a saving if this initial cost is spread over the longer period. Another advantage is that the mineral level of the herbage is usually higher in the older leys. This is due to the fact that there will be greater root development and the plant roots will be searching for nutrient from a greater soil volume.

The amount of root growth has a considerable effect when the ley comes to be broken up. A ley in its fourth year will leave a far greater volume of fibrous root material than one in its second year, and this is valuable organic material for the benefit of ensuing arable crops.

Left to itself nature will always produce a wide assortment of plants in any given situation. In principle the ley should contain as many species as possible, and within the species differing cultivars should be chosen. Some plants will be shallow-rooting, others deep-rooting; some will have an erect habit, others will be procumbent and give a good "bottom" to the sward. Again different species will search out and take differing quantities of minerals, so giving the sward a more balanced mineral composition. Although a large number of species or cultivars will be sown in the mixture, by the time the ley is 3–4 years old dominant species are sure to have taken over and crowded out the weaker varieties, which will have disappeared from the ley. In particular the deep-rooted herbs used in a ley tend not to be persistent. Ribgrass is an exception to this rule.

Opinions differ as to whether the herbs should be sown with the mix or left as a strip down one side of the field. There is an advantage in their being an integral part of the ley covering the entire field, as the benefit of their deep-rooting habit is spread over the whole area. However, it cannot be denied that there is also advantage in having a separate strip of herbs alone where it is possible to grow a much greater variety of herbs and so allow stock a chance to select those which they desire. It is very noticeable how cows will graze out all the "weeds" in hedge bottoms, the banks of the hedges being left quite bare. It is only recently understood how much the cows need the minerals contained in all the weeds.

Perennial rye grass, timothy and the fescues are the most useful of the grasses (in the British Isles) for cattle, and each species has many cultivars adapted for different purposes, grazing or conservation, and so the main use for which the ley is intended must be considered. A ley for sheep will have different requirements from one where cattle are the main concern. There are so many factors which will influence choice of seed mixture (soil type, aspect, height above sea level, results obtained on neighbouring farms) that considerable thought must be given to the cultivars which will be chosen. In general 0.5 kg of white clover per hectare will always be included, although in the first year it may not appear to cover the ground adequately. Some people like to include Italian rye grass and red clover in order that production in the first year will be enhanced. It must be pointed out that as these plants are biennials, when they die out in the second year there is a likelihood that their place may be filled by less productive grasses, such as annual meadow grass, or weeds.

6.2 Establishment of Grassland

The aim of management in the first year is to establish the grass mixture as quickly as possible; after the first year management must be to ensure that the grass stays in a highly productive state for as long as possible.

Sowing is usually carried out by one of two methods:

1. Grass seeds are sown on their own. This method gives the least competition to the young grass seedlings and is the preferred method where the grasses are not very aggressive.

2. More commonly the grass seeds mixture will be sown under a nurse crop either for grain or for grazing. The use of a nurse crop for grain gives the most severe competition but it does allow a cash crop to be taken in the establishment year. The best results, however, are usually obtained when the nurse crop is one that grows quickly, so giving a feed for stock. This grazing should ideally take place 12 weeks after sowing: it is not for animal production purposes but to help establish the grass seedlings. The hooves of the animals will "replant" any seedlings which have "lifted" out of the ground during the germination period. The stocking will be at a very heavy rate but will be for a short period of time, perhaps only half a day. It must never take place when the ground is wet, as the hooves of the grazing animals tear out the young seedlings rather than consolidate them. This short sharp grazing will promote tillering in the young grasses, and therefore more rapid ground cover will occur. Once there is some keep on the field then another short heavy stocking will be required. If weeds become established in the new ley no harm will result from mowing, irrespective of the stage of growth of the young crop, but the ground state must be considered as, yet again, if the ground is wet then damage will be severe. Table 6.1 shows how alterations in the period of time between these early grazings affect grass tillering and weed encroachment. The weeds will for the most part be annuals, and the table also shows the necessity for mowing to prevent seeding. The high tillering rate will ensure that in the second year there is good ground cover, so eliminating annual weed problems.

Young grasses in their seeding year have an ability to continue to grow irrespective of the weather, and in a drought year will provide a great deal of grazing in the dry spells. Towards the end of the first season the rest periods should be lengthened and the severity of the grazing reduced, so allowing the young plants to build up reserves for the coming winter.

TABLE 6.1

Rest period between grazings	Percentage grasses tillering	Proportion of weed to grass in sward
1 month	100	52:100
2 weeks	85	89:100
4 days	70	100:100

6.3   Subsequent Management

Grazing

There are three differing management systems from which to choose:
paddocks, controlled grazing with an electric fence, and set stocking where the
stock are free to range over the whole grazing area. The choice is largely a
personal one although farm geography may influence the final choice.

Paddocks on a dairy farm are usually arranged to give a 3-week cycle so that
at best there should be 21 of them, giving 1 day on each. If this is not possible
14 will give $1\frac{1}{2}$ days on each. Further grazing must be available for later in the
summer when grass growth has slowed down (and it is under this condition that
the young seeds may be important). To have less than 14 paddocks risks
fluctuating yields from the cows, as by the third day the grazing will be inferior.
As a rough guide 0.5 ha for 35–38 cows for 1 day should be sufficient. There
will be considerable expense in laying out paddocks for fencing and water
supply, but once made they are easy to maintain and there is no daily task of
moving fences. Sometimes there will be a considerable waste of land in
providing access to the paddocks. There are further problems associated with
paddock grazing. It is difficult to ensure that they are evenly grazed down in
times of plenty, and it is equally difficult to ensure enough grass in time of
drought or at the end of the season. So the decision of whether to move cows
has often to be faced. There could be too much grass to leave but hardly enough
to give the cows a full day's feed. This dilemma is usually resolved by moving
the cattle on and trimming off the surplus grass with a mower; this is a wasteful
practice. The second difficulty is that, however careful the planning, the day
comes when there is too much grass ahead and one or two paddocks have to be
taken out for hay or silage, and the conserving of grass from these small plots
is a nuisance and expensive.

Controlled grazing with an electric fence is a good system (though the
necessity to move the fence twice a day can be tiresome), as the correct amount
of grass can be made available each grazing period. This must be done carefully
and is a job for the farmer or an experienced stockman. It is as important a job
economically as the feeding of concentrates. To save labour in moving electric
fences it is a good idea to stagger the posts in a zig-zag fashion. Then only every
other post needs moving at each grazing period. Often where the system is
used, so as to avoid heavy treading of the ley, larger fields are divided up by
semi-permanent fences into roughly 4 ha paddocks, and the electric fence is
moved within these "sub-paddocks". At the start of the grazing season about
0.2 ha per cow is allocated for grazing. As other fields are cleared of silage or
hay they can be brought in, so that by the end of July there is about 0.4 ha per
cow available. It should be possible in the British Isles to provide grazing and
fodder for winter from 0.5 ha for each cow. Water is all-important and with a
large herd special provision must be made. After a period of grazing the whole
herd will move off to the drinking trough at the same time. The older cows will

PLATE 9. *Different stocking systems for dairy cows. Above*: Controlled grazing. *Below*: Set stocking.

drink first and unless the tank is very large they will empty it. The rest of the herd have to wait patiently while the trough refills slowly, but all too often when the old cows move away and go to lie down the herd instinct reasserts itself and the younger animals will follow without having had their drink. So three things are called for: large troughs and big pipes to feed them and, most important, enough troughs to ensure that the animals do not have to walk too far for their water. Portable troughs with wheels and handles can be used, keeping the trough level with the cows each time the electric fence is moved.

The problem of water is not so important where beef and young stores are grazed, as their water requirements are not nearly as large as that of the dairy cow.

The third method, set stocking, is a new name for the old-fashioned way. It simply means extensive grazing over the whole of the allotted grazing area. If roads intersect the ground it may be necessary to have two areas – one for the day and one for the night. The amount of grazing allowed under this system is the same as for controlled grazing, 0.2 ha at the start of the grazing season and 0.4 ha per cow towards the end. It is noticeable that the sward under set stocking systems becomes more dense, so allowing less room for weed development, and at the same time the taller grasses which do not tiller as readily as the procumbent species appear to die out more quickly. It has frequently been noticed that when cows are managed under a controlled system they run out to the field after milking, so that they may get their share of the fresh grass, whilst under a set stocking pattern the return of the herd is leisurely, each animal knowing that there will be plenty of grass for all. Certainly set stocked cows appear to be much more satisfied and contented even when there is very little grass to be seen in the field. Another advantage with this system is that the animals can take advantage of the protection of hedges under adverse weather conditions, and it is certainly an advantage not to have to move an electric fence. It does mean, however, that in the autumn mists and darker mornings the farmer may have a longer journey to ensure that no cow has been left behind at milking time. If we may dare to forget economics for a moment, how pleasant it is to see a good herd of cows spread evenly over the landscape instead of in a narrow line behind an electric fence.

Whichever system is used, topping the pasture is most important, and the job can be done with any old mower or forage harvester. With controlled grazing this topping must be carried out as each area is grazed off. At the same time as the topping is carried out the pasture should be chain-harrowed so that droppings are spread and the surface mat is destroyed. These cultivations should ideally be carried out the day after the stock have been moved on. Under pressure of time, chain-harrowing can occasionally be omitted but topping of the pasture is essential; it stimulates bottom leaf production and it is this leaf that the animals require.

There is some conflict as to how soon to start grazing in the spring. For the sake of the pasture the grass should be allowed to get a good start, which will also have the advantage of getting some fibre into it, thus lessening scouring. On the other hand it is better for stock to make the change from winter fodder to grass gradually and this, in an ideal world, would mean turning the cattle out for longer periods each day so that they become accustomed to the young grass as it grows. With adult stock it is usually possible to keep on with dry winter rations when they first go out. Often good oat straw would be available on an *ad-lib.* basis. With calves the situation is different. Often calves are kept in until the weather really warms up. They are then turned out and the sudden change from indoor winter feeding to outdoor grass gives them a most severe check. Some farmers make a practice of sacrificing a field and turn the autumn-born calves onto it during the early spring, but they would still be fed their full winter ration. This procedure allows them to become accustomed to grass very gradually as it grows, without their suffering a check to their growth. Normally these young animals will stand cold and severe weather very well, especially if they are being well fed and have shelter belts to take cover in.

Special Conditions

Early Keep. In the temperate zones of the world after a long hard winter there always appears to be a need to have some grass for ruminants as soon as is possible. The main limit to spring growth is soil temperature. Until the soil has warmed the mycorrhiza and nitrogen-fixing bacteria do not function efficiently, and so any cultural method which will help advance soil temperature is to be welcomed. The most efficient way to get this temperature rise is to have soils with a high organic content which absorb more of the sun's heat, and to have soils with a large pore space, which will result from the use of aerating tools. Besides improving the physical properties of the soil, there are certain grasses which will grow with lower soil temperature. The earliest of all grasses to commence growth is Italian rye grass, followed by the perennial rye grasses. The ideal is therefore to have some pastures which contain Italian rye grass, and indeed under some conditions the establishment of Italian rye grass/red clover leys can be justified.

Dry Summer Gap. A deficiency of grazing occurs when high summer temperatures coincide with low rainfall. Good species to grow to produce keep under these conditions are cocksfoot (*Dactylis glomerata*), sainfoin (*Onobrychis* sp.) or lucerne (*Medicago sativa*) mixtures. It is desirable to keep the pasture well covered with herbage so that soil water loss is kept to a minimum. Under European conditions the ley could be either grazed in May, or rested from grazing during that month taking a light silage crop in early June. In either case aerated slurry or urine would be applied in early June. This gives some growth which will stay green during July when there is little or no growth, and there will be sufficient grass for August grazing. It must also be remembered that young newly sown seeds appear never to stop growing, even in the driest summer, and they therefore can give keep in this difficult mid-summer season.

Bloat. The sowing of legumes in grassland has fallen into disrepute among conventional farmers due to losses in cattle and sheep due to bloat. Bloat is caused by fermentation gases produced in the rumen, in particular by legumes, in such quantities that they cannot escape during the regurgitation process. This accumulation in the rumen causes rapid death. There is evidence, however, that stock which have been reared from animals long accustomed to legumes in their diet are not so susceptible to the condition. This could be due to the young stock obtaining from their environment the same microflora as their parents, the microflora having adjusted to deal with legumes. Notwithstanding the foregoing the appropriate farmer can meet some of the problems by sowing legumes which do not give rise to bloat. The best species are sainfoin and bird's-foot trefoil: both have high tannin contents which take longer to digest in the rumen and so prevent the condition occurring or becoming serious.

Sainfoin (Onobrychis) is well suited to the dry conditions of chalkland or limestones, and is indeed more drought-resistant than lucerne. There are two varieties available: common sainfoin is slow to establish but once established will last for 5 or 6 years in leys; the other common variety, giant sainfoin, establishes rapidly, gives very large quantities of bulk but is only persistent for 2–3 years.

The seed of both varieties has a hard coat, and so as to obtain a quick establishment it is usual to sow "milled" seed, at a rate of 50 kg per hectare. When grown in association with grasses the seed rate would be reduced to about 30 kg per hectare. A typical mixture would be 30 kg of sainfoin, 2 kg of an intermediate type of cocksfoot and 1.5 kg of procumbent white clover.

Sainfoin establishes best when sown in March or April under a nurse crop of oats. Once the cereal has been harvested a light dressing (100 tonnes per hectare) of composted farm manure should be given. Rock phosphates and rock potash would normally be incorporated in the seed bed.

To avoid the crowns being eaten out by sheep, thus killing the plants, it is advisable not to over-graze the crop, and never to allow sheep, which are tight grazers, onto a crop already reduced by cattle.

Bird's-foot trefoil (Lotus corniculatus) is best suited to the light dry sandy soils. It is rarely grown these days and indeed seed is very difficult to obtain. However, where bloat is severe, and where sainfoin will not grow, attempts should be made to obtain seed from one of the speciality seed houses.

Other legumes reported to contain tannins, and so of use where bloat is a problem, are milkvetch (*Astragulus cicer*), hare clover (*Trifolium arvense*) and yellow meadow vetch (*Lathyrus pratense*).

Plant breeders are at present trying to increase the level of tannins in lucerne and white clover so as to reduce the likelihood of bloat occurring when these particular legumes are fed.

Manuring of Grazing Land

The nutrient needs of grazing land are not normally high. The urination of the cattle returns potash to the grass, legumes will provide the nitrogen and the deep-rooted herbs bring up trace minerals in sufficient quantity to ensure that the stock needs are met and surplus is returned to the land via the cattle droppings. The legumes in the pasture will benefit from a light annual dressing of insoluble phosphate fertilizers (basic slag or rock phosphate). If basic slag is used then the lime status prevailing at the time will be maintained, but if the lime level is already low and a great deal of calcium is required it is usually better to lime at a different time in the rotation.

Silage and Hay Production

Winter production of milk or meat from ruminants will depend on the quality of the silage or hay which is available for the stock. No matter what the stock, only the best is good enough, and the choice between silage and hay will depend on local circumstances. On small acreages hay is probably the best proposition, but on larger farms silage or barn-dried hay will be selected. The choice will generally be decided by available capital: silage making or barn hay-drying equipment is very expensive, and a large acreage is needed to justify the capital outlay.

Nothing can really beat good hay, but it is difficult in the fickle European climate to guarantee getting sufficient of top quality.

Silage has several advantages. Grass can be cut at an earlier stage, say at the end of May, than would be possible for hay production. However, it is very important to ensure that the grass has developed some fibrous growth: grass cut before this stage will make superb silage, but without fibre, so that when fed to dairy cows butterfat content of the milk will most likely be reduced. The ideal time to get bulk and quality is just before any flowering heads appear. If a great quantity is to be made then it will be necessary to compromise on time of cutting, or to have leys of different species so as to lengthen the grass conservation season. Silage can be made in dull but dry weather when it would generally not be considered safe to cut hay, but the result will not be good if the grass is gathered to the clamp or silo when it is wet.

Silage loses to hay on the small farm, not only because of the capital costs, but because there is always some wastage on the outside of the clamp and the smaller the clamp the greater the proportion of waste. Silage is a heavy and bulky food to handle and really needs some form of mechanization. Feeding by hand is very hard work, and under-feeding may become a very real possibility.

Silage Making

Any type of mower may be used for cutting the grass. After cutting, the grass should be allowed to wilt for 24 hours so that the moisture content is reduced. The wilted grass can be collected by buckrake, or any other type of forage harvester. However, the best results are obtained if a double-chop machine is used as this gives better compaction in the silo or clamp, and when it comes to feeding a foreloader can work in a double-chopped silage pit without the need to cut out the silage by hand.

The silage heap can be made on the surface of the ground, in a pit or in a clamp under cover at one end of a barn or yard where the stock are to be over-wintered. The positioning adjacent to the cattle housing allows the animals to self-feed during the winter period. The surface heap is the least satisfactory storage method as it is difficult to consolidate and gives rise to more wastage.

When grass is cut for silage some of its life processes continue and it will respire for some time. This means that oxygen is taken in by the plant and some of the nutrients in the plant, mainly sugar, are broken down. While respiration continues large amounts of heat are given off, and this is why the temperature of a heap rises rapidly. Mature grass or cereal legume mixtures do not compact easily and so are more likely to overheat. High temperatures, say greater than 50°C or more, are to be avoided. When silos are used some air is trapped amongst the grass during filling, and this allows excessive respiration causing loss of feeding value; the high temperature also reduces the digestibility of the proteins. The use of double-chopped grass is almost essential in silos. In a pit the best way to prevent undue temperature rise is to run a tractor over the pit, so consolidating the heap. Where a clamp is used this must be covered with a polythene sheet as soon as it has been completed and then made air-tight by weighting. Old tyres are very useful for this purpose, though unsightly.

The latest development in green crop conservation is the use of pure stands of legumes for silage for dairy cows. The resultant silage, when well made, is quite capable of supporting a dairy cow for maintenance and at least 25 litres of milk per day, without the necessity of resorting to dry concentrates. Bloat does not appear to be a problem. This class of silage must be well made, and as there will be only low levels of soluble carbohydrate in the legume, to allow proper fermentation it will be necessary to add some molasses at about 80–100 litres per tonne of green legume. It is usual to mix the molasses with water at the rate of one part molasses to two parts of water, and to spray the resultant mixture onto the, usually, chopped material.

There are now available on the market several types of additive based on bacteria which ensure proper acid formation in the silage. These products show promise in all types of silage making. However, the additives based on formic or proprionic acids, although aiding silage manufacture, are not acceptable as, like chemical fertilizers, they shorten the biological processes with unknown effects on the stock.

Hay Making

Nothing is better than the best hay. However in U.K. conditions it is difficult, if not impossible, to obtain sufficient of the right quality. All too often it is not possible to cut the grass at the right stage, and it does not take much rain seriously to reduce the quality.

Traditional Hay Making. The best hay is made on tripods or wire frames. It takes a great deal of hand labour and is a slow process. The hay is picked up when still quite green, and is put onto the tripods which are about 1.75 m tall. The legs of the tripods are kept secure either with wire or cross-ties about 0.5 and 1 m from the ground. Correctly put onto the tripod, the hay will dry out quite beautifully, with very little waste, and indeed once on the tripod it will withstand a great deal of rain. The biggest hazard is from high winds which could blow the whole tripod over. In Scotland, where the weather tends to be more difficult for hay making, the tripods are much bigger, each holding at least $\frac{1}{2}$ tonne of hay. The success of tripod hay depends on making sure that a good draught of air can come from the bottom up through the hollow central cone. The amount of work involved in this method is high.

If tripods are not used then nowadays most farmers resort to the use of tedders, swath-turning machines and other mechanical aids to help drying, but it must be remembered that every time the drying grass is handled, even by the most delicate of machinery, there is sure to be loss of some leaf, and it is the leaf which contains most of the nutrients. Practically all hay made nowadays is baled either into small square bales or into large round bales which are difficult to handle in the winter feeding period.

Barn Hay Making. For barn hay drying, young leafy grass is cut and distributed onto a slatted barn floor. Air to dry the grass is forced by fan through the stacked grass, which is gradually dried into a very high-quality product. Capital costs for the equipment need not be high, but under wet weather conditions a great deal of air will have to be used to produce the drying (in some extreme cases the air may even be heated) and energy consumption (usually electricity) will be high.

Manuring of Grassland for Silage and Hay Making

Because silage and hay are stored and usually fed to housed animals, the nutrients they contain are removed from the particular field where they once grew. This is the opposite of what happens on grazing land. The loss of phosphate and potassium can be particularly severe where the land is used for continuous hay or silage making (which is not good holistic practice). The nutrient levels reduced by the removed crop must be restored – preferably, where it can be spared, by well-composted manure. The use of fermented urine collected over the winter period has the advantage that not only is the potassium being returned to the fields from whence it came, but the nitrogen content will give the grass an early-season boost. If organic manures are not available then dressings of rock phosphate and rock potash must be given. Under some conditions the pH may have to be restored to nearly neutral by the use of lime. In general liming is best carried out at other stages in the rotation.

6.4   Seeds Mixtures

In appropriate agriculture attempts must be made to obtain an even production of grass all the year round. The original workers in the field, such as Elliot of Clifton Park, sowed a few pounds of many different varieties, some of which we now know have little feed value. However, the concept of keeping the pasture full of different species is still maintained. This wide diversity of varieties not only spreads the peaks of production but also reduces the disease problems, which can be as severe in a single variety sward as in a monoculture of cereals. Sir George Stapledon spent a great part of his life's work selecting, testing and then multiplying different varieties of the same botanical species. For instance in perennial rye grass (*Lolium perenne*) he established three distinct cultivars: (a) S.23 high tillering large leaf with procumbent habit, (b) S.24 a variety which starts growth earlier than S.23 but is more erect and therefore more desirable as a hay or silage grass, and (c) S.101 an intermediate type giving great yield of bulk, but also capable of withstanding grazing. Present seeds mixtures will therefore have differing species of high feed value grasses. The same arguments concerning grasses equally apply to the clovers, which will be interplanted with the grasses so as to give a higher protein leaf. Because of the rhizobium association they also help production of nitrogen in the sward.

Typical Seeds Mixtures

For green manuring the mixture, sown under a cereal nurse crop in the spring, will give stubble grazing in the autumn. After heavy grazing it would be ploughed in during the winter period.

Mixture:

	Italian rye grass	11 kg/ha
either	Broad red clover	2.2 kg/ha
or	Yellow trefoil	2.75 kg/ha

For 1-year ley the mixtures will be sown under a cereal nurse crop in the spring and grazed in the autumn. The following year they will be either grazed or set up for hay or silage, being ploughed in the second autumn. The decision to take hay or to graze in the second year depends on the land fertility. If the land is in good heart then hay or silage may be taken, whilst if the land is in a poor state the pasture would be grazed heavily to keep, and perhaps even increase, fertility.

Mixtures:

1.	Perennial rye grass (cultivar S.24)	15 kg/ha
	Trefoil	1.25 kg/ha, *or*
2.	Italian rye grass	8.8 kg/ha
either	Broad red clover	1.1 kg/ha
or	Late-flowering red clover	2 kg/ha

The broad red clover could be replaced by 3.5 kg/ha of alsike, the latter being more persistent than either of the red clovers and more tolerant of acid conditions.

For 2-year leys: these are very rarely grown, as they necessitate the use of perennial grasses which are only to be used for a short period and are therefore needlessly expensive.

For 3-year or longer leys the emphasis is on long-lived plants. Italian rye grass and the red clovers are too aggressive in the early establishment period and are usually omitted. In designing these long-term leys it is essential that due consideration is given to the selection of the correct varieties of grass and clovers, which will depend both on soil and climate. The ley may be intended for grazing or for general-purpose, but in practice only general-purpose leys would be used in appropriate agriculture. In general-purpose mixtures the following species would be considered:

Perennial rye grass – for maximum production early spring and autumn.

Cocksfoot – more drought-resistant, for early production with a fair aftermath. If cocksfoot is not managed well it tends to become very tussocky, making cultivation difficult. It is deep-rooted and therefore recovers nutrient from the lower soil levels.

Timothy – a very palatable grass but develops late in the season and is not very aggressive.

Meadow fescue – a late producer and not very aggressive.

White clovers – very useful as their rhizobia can fix very high levels of nitrogen. They must be included in all long-term leys.

Red clovers – might be used but tend to be aggressive in the establishment period.

Specialist mixtures for hay or silage only: there is a growing interest in mixtures predominantly for hay, although it must be noted that mixtures of this type will need the addition of urine or potash rocks to restore the potassium removed in the hay.

Mixtures:
Lucerne   27.5 kg/ha drilled at 15 cm,
Timothy S.48   4.5 kg/ha ⎫
Meadow fescue   4.5 kg/ha ⎭ broadcast
Where the land is unsuitable for lucerne the mixture may be altered to the following:
Timothy S.48   11 kg/ha
Meadow fescue S.53   11 kg/ha
White clover S.100   2.25 kg/ha (a good bottom plant).

In the very dry chalklands sainfoin is often sown together with cocksfoot. A typical mixture would be 125 kg unmilled common sainfoin and about 3 kg of cocksfoot S.26. The sainfoin is very drought-resistant and has the added advantage of reducing or even eliminating bloat problems where the field might be grazed after the conservation crop has been taken.

From time to time seed merchants put into their catalogues new varieties of legumes which they consider may offer advantage over older cultivars. This may indeed be true, but the farmer is warned to be certain that these new varieties have the ability to "fix" nitrogen under British conditions. It is possible that varieties imported from different parts of the world may not be suitable hosts for British rhizobia, and inoculation of the seed may be needed.

Deep-Rooted Herbs

The inclusion of deep-rooted herbs in temporary grassland was also pioneered by Elliot, although Arthur Young conducted what might loosely be termed experiments as long ago as 1770. These herbs were included in grass mixtures so that the foliage would provide grazing animals with trace minerals. Elliot had observed that cattle and sheep always ate with relish the herbs found in hedgerows. He later appreciated that the herbs he was using were deep-rooted, which fact had two great advantages: firstly the herbs were recovering plant nutrients which had been leached to levels beyond the reach of many of the grasses – they were therefore recycling nutrients, in particular potassium which can all too frequently become a limiting nutrient; secondly the deep root system helped to keep "open" the lower regions of the soil, so allowing better drainage and therefore aeration.

The most common herbs to be sown with the leys nowadays are:

Chicory (*Cichorium intybus*). This plant is very vigorous, high in mineral content and very palatable to stock. It has a very long taproot which is, as an explorer of lower soil levels, an aerator and nutrient recoverer. The luxuriant growth can be a nuisance when grass is being cut for hay. Additionally if it is not allowed to self-seed it will quickly disappear from the pasture. However the flowering stalk is hard and not eaten by stock, so after the seed has been shed the stem must be topped. It is usually included in grass mixtures at the rate of between 1 and 2 kg per hectare.

Burnet (*Poterium sanguisorba*). A very productive, high mineral content plant. It has a very deep rooting system and thrives on the very dry soils such as those found in chalkland. It is very palatable to sheep. The plant is found commonly on the chalk grasslands, and seed is often contaminated with sainfoin, it being almost impossible to separate the seeds of the two plants by machinery. The usual rate of inclusion in grass mixtures is between 2 and 3 kg per hectare.

Yarrow (*Achillea millefolium*). A very deep-rooting herb containing high levels of many different minerals and palatable to all stock. It is rather difficult to establish but, once estabished, it is very persistent. Up to 1 kg per hectare should be included in grass mixtures.

Sheep's parsley (*Petroselinum sativum*). A very palatable herb which is claimed to have vermifuge properties for sheep. It is not as persistent as most herbs, but where sheep are the main grazing animal the inclusion of about 1 kg per hectare is desirable.

Kidney vetch (*Anthyllis* species). A legume but usually considered as a herb. It is extremely drought-resistant and indeed thrives best on poor dry soils. It has a reputation for not succeeding on good deep loams, and certainly there are better legumes which would be of more value on these better soils. It is usually incorporated in mixtures at 1 kg per hectare.

Ribgrass (*Plantago lanceolata*). The most common herb, narrow-leaved plantain. It is considered by many to be a weed, and indeed it will thrive on the well-trodden grassless areas around water troughs and gateways. If it is included in hay mixtures the leaves will only dry with difficulty, but for silage it would be very acceptable. It is very palatable, very persistent and has a very high mineral content. It is included in grass mixtures at about 1–2 kg per hectare.

6.5 Grassland Improvement

Low-altitude Outrun Ley or Permanent Hay Field

On land at low altitudes grass swards will need improvement because the land has either been understocked, allowing surface mat to develop, or has been continuously used as a "hay field". Because of the set management pattern of taking hay every year, conditions favour the early seeding varieties of grass, which in general have low feed value. These grasses, Yorkshire fog (*Holcus* sp.), soft brome (*Bromus* sp.) and sweet vernal (*Anthoxanthum* sp.) are not very persistent and eventually weeds become established. The land generally will become potash-deficient and before any improvement can be made nutrient status must be restored by use of potash and perhaps phosphate and lime. In both of these conditions the pasture is best ploughed out and brought into the farm rotation. Because the land has been mismanaged for a long period wireworm may be serious, and the first crop after ploughing out may have to be kale, which is wireworm-resistant.

Hill Pasture and Rough Grazings

On this class of land the composition of the sward under British Isles conditions will mainly depend on drainage. If the ground is waterlogged then the *Molinnia* species will thrive, whilst where the drainage is better, or the soil more peaty, *Nardus* species will predominate. On the dry uplands the finer fescues will be the major grasses present.

If cultivation is possible, these grazings can be brought back into a state of relatively high productivity. The gradual decline in the nutrient status of the land due to continuous selling off of store animals must be corrected. Rock phosphate, rock potash and generally lime will be required, and these fertilizers should be applied in the year before reclamation is started. The most rapid method of obtaining improvement in the pastures is by the use of a soil reconditioning unit. One pass of the tool will introduce air and allow of better soil microbiological life. If the land is wet then reverse Keyline practices are an ideal cultivation method. If a seeder is used at the same time as the tool better-quality grasses and legumes can be sown. It could be well worth sowing a pioneer crop before the re-establishment of grass. A seed mixture consisting of 2.5 kg grazing rape, 2.5 kg turnips and 10.5 kg Italian rye grass per hectare is often sown. The pioneer crop is followed then by a well-balanced grass mixture which would in all probability be sown at the same time as a further cultivation with the soil reconditioning unit is being made.

As the land gradually improves, the use of the soil reconditioning unit will not be required every year, but mat will still develop and it is essential that these re-established lands are sytematically harrowed to remove mat and always topped with a mower. Topping is essential to prevent the early-seeding grasses yet again becoming established. Occasional applications of rock phosphate and rock potash will be needed.

Bracken often becomes a nuisance where no mechanical cultivations are possible. Under these conditions it is best to cut the bracken (if necessary by hand) in June and August, and then plant trees in the autumn.

Allelochemicals have recently been discovered in bracken[1-3] which are harmful to some grasses; therefore to attempt to restore the land to grass, even if cultivable, would prove difficult.

References

1. Gliessman, S. R. (1976) Allelopathy in a broad spectrum of environments as illustrated by bracken. *J. Linn. Soc., Bot.*, **73**, 95–104.
2. Gliessman, S. R. and Muller, C. H. (1972) The phytotoxic potential of bracken. *Kuhn. Madrono*, **21**, 299–304.
3. Gliessman, S. R. and Muller, C. H. (1978) The allelopathic mechanisms of dominance in bracken in Southern California. *J. Chem. Ecol.*, **4**, 337–362.

Chapter 7

Animal Husbandry

IN holistic agriculture animals are an essential part of the whole farming enterprise and should not be considered as separate individual units. In general stocking rates will be at a lower level than that experienced in the modern livestock units, and housing systems, where needed, will be less costly per animal housed. Growth-promoters, drugs for supra-ovulation, methods of artificial rearing from an early age, all of which are considered essential in conventional agriculture to reduce the building cost per animal reared, will not be used. The lower stocking rate will reduce the rapid spread of disease so often seen in intensive units, and the need for a high veterinary medicine input in the form of vaccine and drugs administered at curative or preventative levels will disappear.

The system adopted for housing will be designed not only with the animal's welfare, but also the profitability of the whole farm, being considered. Livestock units, besides adding to farm income, should be designed to allow the conversion of straw into farmyard manure for composting, and hence adding organic matter to the soil and general nutrient to the crops, and in particular as a method of distributing potassium to those crops which have a high demand for this nutrient. The ruminants convert farm crops such as grass and straw, not directly usable by man, into human foods (milk, milk products and meat, etc.) and clothing (wool, hide and mohair).

It is inefficient to use cereal grains which can be used for human consumption as a source of food for pigs and hens to make meat and eggs; therefore these monogastrics will be fed on the by-products of the human food industry (skim milk, breakfast cereal waste, brewers' grains, meat and bone meals, household swill, bakery waste and the residues from the crushing of oil seeds, citrus and olives). As well as this almost endless list there are also the farm's own by-products – tail corn, small potatoes, rejected carrots, pea haulms, unmarketable broccoli – which can all find a place in farm animal nutrition. Crops such as sweet lupin, beans, field peas and fodder beet may also be grown as part of the farm's rotational cropping system.

7.1 Monogastrics

Poultry

There is no room in holistic agriculture for the most intensive systems of housing, such as those found in battery cages for egg production or broiler houses for the poultry meat industry. These systems have grown in popularity, as it is believed that intensity leads to lower costs of production. However, it has been shown that in general as long as the diets for the hens and the period of lighting is kept the same as that used in battery systems then production and costs are similar. The following has been quoted verbatim from the AFRC Poultry Research Centre Report, December 1984 (Abstract 80).

ETHOLOGY, HUSBANDRY.

Title: A comparison of laying stock: housed intensively in cages and outside on range. Years 1981–1983.

Authors: Hughes, B. O. and Dun, P.

Publication year: 1984.

Reference: WSAC Research and Development Publications No. 18.

SUMMARY

This study continues a series of trials comparing the performance of laying hens housed either in battery cages or outside on range.

The birds (ISA Warrens) were reared on litter to 18 weeks of age and allocated at random: 300 to cages (48 × 41 cm) in groups of 4, and 152 to four outside pens (23 × 21 cm) in groups of 38. Two pens had accommodated birds for the previous two years (old pasture); the others had not supported poultry recently (new pasture).

Hen-housed egg production over 52 weeks in cages was 280, and on old and new pastures 286 and 280 respectively. Both mean egg weight and feed intake were greater on range (61.5 g, 140 g/day) than in cages (59.7 g; 119 g/day). The apparent efficiency of food utilization was about 15% better in cages (0.401) than on range (0.348). The true figure on range was even lower, because consumption of grass was appreciable; perhaps as much as 30 g of dry matter/bird/day.

Body weight increased more rapidly on range, averaging 2.35 kg at the end of lay, in contrast to 2.19 kg in cages. Feather condition was initially very good in all birds; it changed little on range but deteriorated markedly in cages. At the end of the study all parts of the alimentary tract were heavier in the birds on range. This is consistent with the consumption of more feed with a higher fibre content.

The proportion of cracked eggs was initially low (1 to 3%) and similar in both systems. It remained low on range but in cages rose to over 10% by the end of lay.

The winter was extremely cold (−18°C outside and −13°C inside the hut), without having any apparent adverse effect on production.

This environment was one which allowed the birds to express a wide range of behavioural patterns. No evidence of infectious disease or parasitism was encountered, but the lack of any disease up to now does not mean that the danger is not constantly present in a system of this kind.

In Abstract 79 of the same publication, given below, the same authors, apparently reporting the result of part of the same experiment but with the results of more than one year included, give more details of the management of the range birds.

ETHOLOGY, HUSBANDRY.
Title: Production and behaviour of laying domestic fowls in outside pens.
Authors: Hughes, B. O. and Dun, P.
Publication year: 1983/84.
Reference: *Appl. Anim. Ethol.*, 11:201.

SUMMARY

Egg production in extensive systems, is generally thought to be lower than in intensive systems such as cages (typically 190–210 eggs *per annum* rather than 260–280). For this and other reasons it is argued that modern hybrids are particularly well adapted to caging.

In this small scale controlled trial (see also Poultry Research Centre Report for the year ended 31st March 1981, abstract 39) two groups of 38 hens were housed in outside pens, 23 × 21 m, and compared over a 20-month laying period with controls in groups of four cages. They were fed the same food, inside their hut, and almost all eggs were laid in inside nest boxes. Supplementary light was provided before the natural dawn inside the hut so that the photoperiod was at least 14 hours.

The birds outside produced as many eggs (337 in 560 days) as those inside (378), and they were nearly 2 g heavier. They ate more food: 135 g/day as opposed to 121 g/day, and supplemented their diet with grass, (up to 30 g/hen/day). Mortality was similar in both systems: 10.8% in cages and 11% outside. The birds in pens were heavier and their feather condition was better.

Outside access was provided between 0800 and 1700 hours; the number outside being dependent on weather conditions, averaging for the medium hybrids about 47% in dry, sunny weather and 31% when wet and windy. For light hybrids all figures were 30% and 21% respectively. The numbers were variable, with a relatively constant high rate of movement. The number emerging within five minutes after the pop-hole was opened at 0800 hours was variable (range of 6–83%) from day do day, but the two flocks showed a high correlation, suggesting some common environmental factor. No simple relationship with weather conditions could explain this variation.

The main activities outside were grazing, ground-pecking, ground-scratching and dust-bathing. The vegetation was heavily worn near the hut, but birds rarely entered the peripheral areas, which were essentially ungrazed. There was no evidence, from production or mortality data, that modern hybrids are specifically adapted to either intensive or extensive systems. The behavioural observations suggest that a full repertoire of foraging behaviour is still present.

Chickens

Day-old chicks not reared under broody hens will need artificial heat for the first weeks of life. This heat may be given from above with infra-red lamps or from electric under-floor heating coils or, more traditionally, from well-insulated areas where the initial heat is provided from a chimney to a kerosene light.

Growers can be kept in housing systems similar to that described in the subsection dealing with laying hens. Indeed it is advantageous for the young growers to be reared in the same pens as will be used for laying accommodation.

Layers: methods have been developed for keeping hens economically on systems which are considerably less intensive than hen batteries. Figure 7.1 shows one suggested layout developed by Lady Eve Balfour in about 1948.

Five square metres of grass, preferably temporary, is allocated to each hen. This total grass area is divided into two equal sized paddocks, D and E. Access for the hens into one of the two paddocks from the straw yard (B) is by way of the pop-hole, and the gate is swung to position G or G'. The paddocks must never become muddy patches, and as soon as they show damage they should be shut up and the hens allowed access to the other paddock by moving the position of gate G. Ideally the paddocks should then be topped with a mower and not used again until they have produced about 5–7.5 cm of new grass growth. They must not be used for more than 2 years, and new paddocks should then be erected at D' and E', or the whole site moved to new temporary grassland. "Free range" hens are often housed on small grass areas and are condemned by their owners to live in quagmires. This type of husbandry is as much anathema to the holistic farmer as are battery cages.

Adjoining these grass paddocks is a straw yard, also allowing 2.5 m^2 per hen so that each laying hen has a total of 5 m^2 of exercise area at any one time. This is a scratch area for the birds, and it is here that they will receive, last thing at night, a scratch feed of grain. Straw is continuously added so that the hens can continue to scratch and turn the straw into good-quality compost. The open yard is cleared once a year and the material removed can be further composted. As has already been discussed in the chapter on plant nutrient, there is no risk in using thoroughly composted chicken manure, but uncomposted material is over-rich in ammonium ions and can cause scorch and upset the mychorriza and the ethylene cycle. Good composting will ensure any parasite eggs are destroyed. Covered feeders or troughs are placed in the straw yard to hold the meal which the hens receive on an *ad-lib.* basis. Trough space should be allocated at about 11 m of length for every 100 birds (5.5 m of double-sided

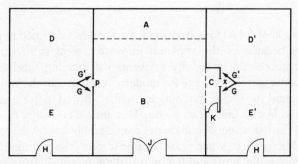

FIG. 7.1   Suggested layout for Hen yards used in the Balfour system. A: Open fronted shed, with perch length allowing 9″ per bird; B: deep litter straw yard; C: laying house with dust bath under; D/D′: grass paddocks; E/E′: alternative grass paddocks; G: gate which can be swung to alternate position G′; H: gate to allow access for mowing; J: gate to allow access for tractor clearing out of straw yard on an annual basis; P: pop-hole allows access from straw yard to either paddock D or E; K: access door to laying house; X: pop-hole to alternative paddocks from the dust bath area.

trough). Water is best supplied through an automatic supply, 1.5 m of trough for 100 birds. If the pipe supplying the trough is flexible then the trough can be moved around the yard, so ensuring that no area is left to become a quagmire. A dust bath should also be provided, and this is often the space left below the laying house (C) adjoining the open straw yard and containing nest boxes. Nest boxes should be about 2 m long, 60 cm wide and preferably with a wire floor kept covered with a thick layer of straw. Entrance is through a pop-hole, 20 × 20 cm. The roof of the nest box is hinged to allow access to collect eggs. It is usual to have a sloping roof to stop hens perching on top. The box should be about 35 cm high at the front and 75 cm at the rear. A nest box of the above dimensions will be sufficient for up to 60 hens. The length should of course be increased proportionately if more than 60 hens are kept.

Table birds are satisfactorily reared on the same system as that outlined for layers.

Ducks

Different types of ducks can each have a part to play in holistic agriculture, e.g. Khaki Campbell for egg production and the large traditional Aylesbury for meat. In addition to their specific purpose of egg production or table meat, they all have one very great advantage in that they will eat the water snail, which is the secondary host to sheep liver fluke, and therefore where sheep are kept the duck is an almost essential requirement of the farming system. They are not difficult creatures to rear; a few days after hatching they will be eating well, and if water is available will already be exercising. To control the snail they must obviously be moved to those wet areas where the snail will be found. Housing at night is required to protect ducks from the fox – almost its worst enemy.

Geese

The goose, kept for so long by British farmers, has declined in numbers in recent years because of the apparent impossibility of subjecting it to the intensive methods demanded by conventional farming, and because the Christmas goose can be too large for modern ovens. Again the young are very easy to rear, and the spring hatching is usually allowed to graze and produce the traditional Michaelmas grass goose. This animal is smaller in weight than the fattened Christmas goose and carries considerably less fat than the finished bird. After Michaelmas, when grass became scarcer, the goslings were usually penned in deep-litter yards and fed on cereal mashes made from tail corn and household scraps to fatten for the Christmas trade. The goose will produce, under barbaric conditions, fois gras, the gourmet's delight, but if the goose is kept humanely its liver can still provide the purchaser with a delicious raw material for his paté. Goose grease – the basis of many holistic remedies – and down can also be utilized.

Turkeys

The turkey is the most difficult of all poultry to rear in a holistic manner. The disease "blackhead" can wipe out whole flocks. Conventional farmers overcome the problem by the continuous feeding of drugs which give total protection and so allow the intensification of turkey rearing and fattening. These drugs are in no way acceptable to the holistic farmer, and management systems have to be developed to keep the turkey poult away from the causative organism found either in turkey droppings or in some badly infected land. Before the advent of drugs turkeys were reared on wire floors and the droppings were pushed through the wire by the birds treading, and so contact was broken. These rearing methods are not much, if at all, better than those involved in continuous drug feeding, as the damage to the feet was horrifying. The only real alternative is to rear the turkeys on fresh grassland in types of "arks" which can be moved to new ground each day. As soon as the land becomes too wet and the turkeys need housing for "finishing", the birds can be moved into covered yards, but with plenty of light and air, and be kept on a good depth of chopped straw. They can scratch in these yards and so help spread their droppings which will compost, and the composting process will ensure that the disease organisms are destroyed.

Pigs

In holistic agriculture animals are considered as part of the whole farm and national enterprise, and their justification is in their ability either to use products not available to man as food, or to convert farm waste products such as straw into fertilizers, which in themselves are soil conditioners.

The pig is capable of using waste human food, and because it is virtually hairless has a high demand for bedding straw. Because of both demands it makes an admirable animal for holistic agriculture. No wonder the Chinese consider it to be a walking compost machine.

Sows and Litters

Sows should where possible be allowed to run out on grassland both during the pregnant and suckling period. In the Roadnight system the only housing required is a small half-round corrugated hut with a wooden floor and an adequate supply of bedding straw for each sow. The grazing will meet practically all the fibre needs of the sow, as well as some of the other macro-nutrients and all the micro-nutrients. Due to the extensive nature of the husbandry system the feeding of "wet" foods is not really practical, but large "sow cobs", made from dry waste products, scattered over a large area, are practical. A plentiful supply of water is needed, and in winter conditions it is necessary to ensure that the water supply has not frozen, so depriving the herd of its water. On many units plastic water pipes are mole-ploughed into the field. Where the pipe comes above ground to the drinker, manure is placed over the exposed pipe, the heat generated by the composting process being sufficient to keep the pipe unfrozen. It is usual to pen 20 dry sows and a catch boar in a 1 hectare paddock fenced by two strands of electric wire. About 3 weeks before farrowing the sows are divided into two groups of 10, and this number of sows with their litters will also occupy a 1 hectare paddock. Roadnight, when he started the system, farrowed all sows at one time, but now with this outdoor system it is usual to batch sows so that one paddock will farrow within a short period of time. This batching gives a better cash flow and more even farm labour usage.

At farrowing sows kept under these conditions will take their own hut, make a nest by chewing up bedding straw, and produce their young with no difficulty and very little loss. Once the piglets are about 3 weeks of age they must be given access to creeps – areas to which the sows have no access but where the young piglets can obtain highly palatable foods which will make up the nutrient demand when the sow's milk supply is starting to decline. Again the food should consist of mixtures of rejected human foods – energy can be supplied from the rejects of the breakfast cereal market, or the farm's own tail corn, whilst high-quality protein can be obtained from fish or meat and bone meals. The by-products which are derived from the human food industry – whey, waste potato crisps, swill – very frequently have a relatively high salt content. Farmers who use these products should ensure that the vitamin mineral supplement does not contain salt. Sufficient is contained in the raw material, and any addition could incur a real risk of salt poisoning.

It of course goes without saying that the sow herd will rotate with the crops round the farm; therefore their urine and faeces will be spread to the fields in rotation. As sows can damage grassland with excessive rooting, they are often allowed the last year of a grass ley – the added nutrient will be of benefit to the following winter wheat crop. If grass has to be protected from the sows rooting then ringing may be considered.

Where land is not dry enough to carry sows and litters out of doors, the sows should be kept in deeply strawed yards, with boars housed in pens between the yards to check oestrus. The yards should be fitted with individual feeders so that every animal obtains its correct food allowance. When sows are yarded the feeding of wet waste feeds instead of dry sow cobs becomes feasible. In a new feeding arrangement sows are given collars fitted with radio transmitters which are specific to them. When they wish to feed they enter the feeding bay and a quantity of food is automatically dropped into the trough. The amount dropped is recorded by a computer. The sow may leave the stall at any time and is free to return at any time to eat up her daily limit. Once the limit has been reached the sow's radio transmitter activates the computer, which notes that the animal has had its correct quantity of feed and does not allow it further access. It might appear to the reader that the method is cruel, but in actual fact closed circuit monitoring of sows kept in this manner shows that very rarely do they want to feed more than once or twice a day, so they do not become frustrated. More importantly the daily printout by the computer of the amounts of food eaten by each sow alerts the stockman to seek out those animals not eating. Animals which do not eat properly are usually in the first stage of illness, and so animal health checking under this system is much upgraded.

About 7 days before farrowing sows are moved to farrowing quarters where they can produce and rear their young. Rails should be provided to prevent piglets being crushed against the pen walls. The administration of organic iron salts to each piglet, or the provision of freshly cut turf in each pen, will prevent the occurrence of iron deficiency anaemia in the young piglets. Very recent studies in America have noted that young piglets spend a great deal of time rooting in sows' faeces, and that if the sows are therefore fed on a diet with a high iron content then anaemia is not seen in the young piglets. The use of the modern injectable iron preparation is expensive, and dirty needles can give rise to abscesses in the finished pig carcase. In the farrowing houses there is more air space than in the outside small farrowing huts; it is therefore very difficult, if not impossible, to keep the room temperature high enough without provision of a supplementary heat source. Usually creeps are fitted into the farrowing pens and creeps can be heated with either infra-red lamps or liquid petroleum gas heaters, which will both attract the piglets and keep them warm. Obviously the creep area can be used as a place where supplementary feed can be offered.

Age of Farrowing and Weaning. Over recent years there has been a trend to farrow sows with their first litters at 10–12 months of age, and to wean the piglets routinely at 2–3 weeks of age. Such methods certainly stunt the growth of the sows, and in all probability restrict the active breeding life of the sow.[1,2] The very latest research work[3] shows that sows which are born out of doors have a longer breeding life than those born under intensive conditions.

The weaning of piglets at 2–3 weeks undoubtedly leads to an increase in housing and feeding costs for the young pigs, and it appears that more and more powerful drugs are required so that disease may be kept to acceptable levels. Again with very early weaning it appears to be more difficult to get sows back into pig. The holistic farmer would in all probability not farrow gilts until they are 13–14 months old, and would not wean piglets before they were at least 5 weeks old and more likely 6–7 weeks of age. At this older age the piglets are eating well, and when they are housed for finishing they will not need expensive follow-on housing and post-weaning checks will be minimal. It will be argued that unless each sow produces about $2\frac{1}{2}$ litters per year she does not produce enough pigs to meet costs. This is a typical reductionist argument where only the pig enterprise is considered. No thought is given to the fact that holistically fed pigs will have far lower feeding costs, housing costs will be minimal, fertilizer costs reduced, and soil condition improved. When all the benefits are considered the holistic approach will always be the most profitable.

Finishing Pigs

The ideal housing for growing and finishing pigs is in yards with plenty of bedding straw, and for younger pigs it is beneficial to have a low kennel where they can sleep out of any draughts. If the kennel is kept low in height then straw for bedding the yards can be stored above, and it is then very easy to throw down the straw into the yard as it is required.

The pigs when weaned should be mixed together into groups of between 10 and 20 pigs of the same weight. They should be allowed at this stage a total of about 0.77 m² of floor area, but if it is intended to keep them in the same pens until they are finished at a heavier weight then about 1.25 m² will be needed. It is good practice to house a bunch of say 20 weaner pigs in a pen, and then at a later stage to split them into two groups of about the same size. In this way unevenness and consequent bullying is avoided, and pen sizes can all be of the same dimensions. Tail biting,[4] which can at times be a problem, is not solved by docking the pigs' tails, but by providing the pigs with something to exercise their minds – straw bedding to chew, or chains dangling from the roof with which they can play. The good husbandman will watch his pigs at feeding time and make sure all the pigs in a pen come to feed. A pig not feeding is generally a sick pig, and will probably need attention.

After weaning, bulky waste products such as swill, small potatoes and vegetable waste should be fed. It is essential that adequate trough room is allowed – about 30 cm per pig housed will be needed.

It cannot be overstressed that when swill is fed to pigs the food must be boiled and treated as required by the appropriate veterinary authorities.[5] These regulations are imposed to control the spread of diseases such as swine vesicular disease, and any farmer who does not comply with the regulations within both the letter and spirit of the law is not considering the holistic approach.

7.2 Ruminants

The ruminant animals stand supreme in agriculture in that they alone are capable of converting plant cellulose into meat, milk and so dairy products, hides and fibres for shoes and cloth.

Modern agriculture does not appear to have made much advance in the intensification of the ruminants, although the battery rearing of calves for veal, and the more recent introduction of the clipping of sheep in the winter and then housing them intensively during the winter months, are notable steps down the reductionist path.

Cattle

Calves should, at a minimum, be given sufficient of their dams' colostral milk to ensure that they receive adequate immunological protection. It is usual on most farms for beef calves to be suckled for 6–9 months, and for the beef herds to spend their whole life out on grazing land. The store cattle produced in these beef herds may be finished off on grass, or perhaps sold to lowland farmers who would finish the beasts in covered yards, similar to those that would be used for over-wintering dairy cattle. Holistic agriculture would not be concerned with any management in the fattening period which did not allow full rumen use. Barley beef, where it is policy to reduce fibre intake, or the more recent systems devised where cattle are being made to drink large quantities of liquid feeds to meet their nutrient requirement, are totally unacceptable.

The dairy calf is best reared under a multi-suckling system. In this system calves are fostered on to cows which are difficult to milk by machine, or who perhaps have lost a quarter by accident or disease. The number of calves fostered at any one time will vary according to the milk yield of the foster mother at the time. The use of substitute milk powders based on dry skim milk and vegetable oils is possibly acceptable under very special conditions – such as those ruling in countries where fresh milk for human need is in critically short supply. Some of the "calf milk" substitutes which are now available on the market are acidified, and are totally unacceptable to the holistic farmer. The acidification does not provide any special nutrient for the calf, and indeed may inhibit proper rumen development or natural rumen function. The acidification is carried out by the manufacturer to ensure that the product when made up into a liquid can be used through an automatic calf-feeding machine without clogging up the feeding tubes. The promoters of these systems claim that their use does away with the need for weekend supervision of the animals. This may be so, but no holistic farmer would ever contemplate leaving any animal, let alone young calves, for 36–48 hours without observation.

The ideas behind the use of acidified calf milks shows up in sharp relief a typical difference between conventional and holistic farmers.

In general it is desirable for a calf to receive at least 450 litres of whole milk before weaning on to highly palatable foods, designed to change its digestion from the monogastric calf to the fully developed ruminant. Housing must always be light, airy, and with a plentiful supply of good bedding. Once the dairy calf has reached the age of 6 months it is able to go out to grass, and if the grazing is poor supplementary feeding will be required. Some primitive shelter should be provided in the winter periods. The old idea of rearing calves in the "calf paddock", a field of grass near the buildings, is to be deprecated. The pasture becomes infested with parasites, and only sickly calves in need of veterinary treatment will be reared. The rotation of ground in the prevention of animal disease is just as important as rotations in prevention of crop disease.

As has already been stated, the housing of calves in wooden crates and feeding them on iron-deficient diets so that white meat carcases for veal can be produced is anathema, and must never become a part of holistic agriculture.

Normally dairy cattle will be given access to grass in the summer months. The management system of grassland for them has already been discussed in an earlier chapter. During adverse winter conditions dairy cattle should be kept in well-strawed yards, usually with a centre passage to allow silage or other roughages to be brought in. Sometimes the yard adjoins silage pits where animals can serve themselves to the roughage. The yards should be arranged so that the cattle can be easily controlled for milking through normal milking parlours where further rationed food can be provided. The mucking out of the yards will usually be at the end of the winter period, but if need be it can be carried out during the winter season. However the heat developed as the straw is decomposed does give a warm bed on which the animals can lie.

In some countries where animal protein is in short supply for fattening pigs, it is often policy to run pigs in the yards with the cattle. The rumen bacteria produce vitamin B_{12}, an essential for monogastrics, which is usually found in animal protein, but the pigs in the yards obtain their supply of the vitamin from the cattle faeces contained in the bedding where the pigs will root. The space required for cattle housed in yards is not less than 8 m^2 per animal housed. If the cattle are to be left with horns then it is vital that they are given at least 50% more room, so that they do not damage one another. It is not usual to group yarded cattle into batches of more than 15 animals.

On dry land conditions such as the chalk downs the "Hosier" outdoor milking bail is ideal, and an excellent system, although not as comfortable for the dairyman on a cold winter morning. Under this system it is the milking bail which is moved round the fields, and not the cows moved twice a day to a stationary milking parlour.

PLATE 10. *Outdoor housing of breeding stock. Above*: Sow and litters on the Roadnight system. Note 'playpen' for very young pigs (*Farmers Weekly*). *Below*: Hosier milking bail (Hosiers, Ltd).

Sheep

Sheep are of great importance to the holistic farmer as they convert grass into saleable products and so help the farm's gross income. Additionally as they are very tight grazers, they make a very good alternative to the cow as a grazing animal. This alternative grazing ensures that grassland is more easily kept in good condition.

There is a multitude of differing breeds and crosses of sheep, and which particular breed or cross is to be kept will depend on each individual farm, wet or dry land, cold or hot climate. Market forces will determine whether quality wool, early lamb or a combination of both are required, and again this will have an influence on the breed which is to be kept.

In general sheep should always be kept out of doors and their requirements are simply those of good husbandry, i.e. foot paring, dipping, clipping, and regular observation for signs of illness. At lambing time a fold should be erected to allow protection for the ewe and offspring for its first few days of life. As soon as the lambs are sturdy enough they should, together with their mothers, be moved to fields where there is shelter in inclement weather. The date for the commencement of lambing should be fixed so that in normal years, weather conditions and food availability (grass) is correct. The arrangement of mating so as to catch a speculative "early" lamb market seems unreasonable, and usually leads down the road of intensification with all its consequential problems. The early breeding of ewe lambs, like pigs, seems to be detrimental to future productivity and would therefore appear to be self-defeating.

Sometimes ewes are clipped and housed overwinter to prevent damage to land, so wet at this time of year that great damage would be caused to soil structure. This would appear, on first consideration to be acceptable, and indeed preferable to the flock being kept under atrociously wet conditions. However, the question can be asked, would not summer-fed fat lamb production be more appropriate on this heavy wet land? Each farmer must answer the question to his own satisfaction. What is certain is that the housing of the ewe for a welfare consideration is acceptable, whereas housing to satisfy intensification is not acceptable.

Goats

Goats have perhaps the worst image of any of the farm animals. They are not to be dismissed lightly, however, as they can offer an alternative to the cow where land is in short supply or not good enough to support a dairy herd.

It is not normally the practice to keep goats on a large herd basis; more often they will be tethered so that they might graze or browse on waste banks or roadside verges.

If a herd is kept it will require about 0.1 ha per goat per year for grazing and about 0.05 ha of grass for hay production. Set stocking is not satisfactory for goats as they will tend to overgraze parts of the paddock whilst allowing other areas of the grass to run rank. It is therefore most usual to graze rotationally, so ensuring that the grass is managed properly. Some goat farmers keep their animals housed all the year round and will zero-graze the land, i.e. freshly mown grass is carried to the housed goats each day. With this system the goat has no freedom, and although grassland is kept in perfect condition it is not an acceptable system when welfare is considered. Goats, unlike other domesticated ruminants, are more a browsing animal and can make good use of areas of land which are out of control – blackberry thickets for example. If they are tethered they will kill out the blackberry and are therefore very good as pioneer or "reclaiming" animals. They utilize muka (especially willow muka) very efficiently. Where they are used to clear blackberry they require about 0.4 ha of scrub per annum.

Under U.K. conditions goats will need to be housed during the winter period. There are three housing systems which can be considered. The first is a stall system where the goat is tethered in the same way as cattle. The animal will receive all its nutrient in the stall where it is being overwintered. Under this system, which is not as acceptable as other systems, each goat should be allowed about 1.5 m² of covered space.

In the second system, perhaps the ideal but also the most expensive, the goat is loose-housed in a pen of at least 2.8 m² in size. The animal has complete freedom and there is no chance of bullying. The last system, often called the dormitory system, is in reality housing in covered yards. Each goat is allowed about 1.4 m² of floor space and has complete freedom of movement. Milking by hand or machine can be carried out in a machine milking bail or in a separate milking room.

References

1. Ukhtverov, M. P. *et al.* (1985) Extending the breeding life of sows. Abst. 1314, *Pig News and Information*, **6**(2).
2. Harog L. A. den (1983) Access to pasture for sows. Abstract 1578, *Pig News and Information*, **4**(3).
3. Dyck G. W. *et al.* (1985) Effect of gestation housing on reproductive performance of Yorkshire and Yorkshire × Lacombe Sows. *Can. J. Anim. Sci.*, **65**, 221–9.
4. Kudryavtsev, A. (1983) Housing density and aggression in pigs. Abstract 2256, *Pig News and Information*, **4**(4).
5. Anon (1984) Swill Feeding Booklet. Ministry of Agriculture, Fisheries and Food.

Chapter 8

Animal Nutrition

HOLISTIC livestock husbandry, like holistic crop husbandry, does not decry yield (except where it is pushed beyond that which is sustainable or becomes reliant on drugs, or where the welfare of the animal is not considered). Like crops animals have the same nutritional needs irrespective of whether they are farmed conventionally or appropriately, but in holistic agriculture the animal will where possible be fed on one or more of the multitude of by-products which are available, and which are generally today wasted, or on crops which have little or no value to man, but which form an essential part of the holistic farming system.

In Appendix I a list of the nutrient make-up of the common feedstuffs is given, and the use of this information should allow for proper formulation of diets. However, many foods on offer, and which can be used by the holistic farmer, may not have published analyses. Therefore before they can be considered for use in a diet for any particular class of stock, chemical analysis for oil, crude protein, crude fibre, ash, moisture, calcium and phosphorus will be needed. From this analysis it is often possible to estimate the available lysine and other amino acid levels of the feed, by comparing it with a similar proteinaceous food. For instance there will be little difference in the amino acid make-up of a 50% protein meat and bone meal or a 45% protein meat and bone meal; therefore it is permissible to extrapolate, given the values for one, the content of the other.

Energy values will have to be determined, and for the monogastric hen and pig the following formula has been shown to give good estimates of the digestible energy of the food measured in megajoules per kilogram.

Digestible energy = 0.275 × Percentage oil content of the food
+ 0.201 × Percentage crude protein content of the food
+ 0.017 × Percentage crude fibre content of the food
+ 0.157 × Percentage nitrogen-free extract content of the food

The nitrogen free extract is the term given to the remainder of a food when the percentage moisture, oil, crude protein, crude fibre, and ash is deducted from 100.

By definition 100 total digestible nutrients (shortened to TDN) used in pig diets are 18.4 megajoules of digestible energy; therefore the TDN is calculated by multiplying the digestible energy value by 5.435.

For hens, where the energy values are more often expressed as metabolizable energy (expressed in megajoules per kilogram) the digestible energy calculated for pigs can be used as a very good approximation. For example: consider a by-product of the British brewery industry which has the following chemical analysis:

15% moisture
3% oil
8% crude protein
7% crude fibre
15% ash.

It is desired to find the approximate content of available lysine, methionine, digestible energy and TDN.

From analysis tables we learn that barley (and it is probable that a British brewery waste product will be based on barley) has a protein level of 10.9%. The available lysine and methionine levels are 0.32% and 0.17% respectively.

Therefore by proportion the product we are considering will have an approximate available lysine level of:

$$0.32 \div 10.9 \times 8 = 0.23$$

approximate methionine level of:

$$0.17 \div 10.9 \times 8 = 0.12.$$

The nitrogen-free extract for the product is

$$100 - (15 + 3 + 8 + 7 + 15) = 100 - 48 = 52.$$

The digestible energy equals

$$\begin{aligned}
0.275 &\times 3 \text{ (oil)} + 0.201 \times 8 \text{ (crude protein)} \\
&+ 0.017 \times 7 \text{ (crude fibre)} \\
&+ 0.157 \times 52 \text{ (nitrogen-free extract)} \\
&= 10.87 \text{ MJ/kg}
\end{aligned}$$

The total digestible nutrient equals

$$10.87 \times 5.435 = 59.08, \text{ say } 59.$$

These calculated values are on a dry matter basis and where the product appears "dry", like cereals and oil seed cakes, they can be used as calculated. However, when the product is obviously liquid (such as household swill) or is known to contain a very high proportion of moisture (raw potatoes or fresh vegetables, for instance) it is usual to alter the results which have been calculated on a dry matter basis, to a "meal equivalent" basis. Meal equivalent has been arbitrarily fixed at 85% of the dry matter value. This 15% diminution is considered as the moisture content of the apparently dry foods. Therefore when diets are computed all raw materials are considered on the same basis.

Nutrient specifications for various classes of livestock are given in Appendix II.

It is perfectly straightforward to formulate diets to the nutrient standards suggested, but when a large number of nutrient requirements have to be met the calculations become tedious. The mathematics are further complicated if "least-cost" formulation is considered. Least-cost formulation is, as its name implies, the calculation of a formula for a diet, of given specification, from various raw materials which have differing nutrient content and differing costs, to the lowest cost. The method is widely used in conventional agriculture, and indeed in industry where blending of raw materials to a specification (for instance dry soup mixes) is important.

The advent of the computer has made least-cost formulation very rapid and therefore it has increased the scope for considering more nutrient standards and more raw materials in a diet. This enlargement of the formulation base, in terms of nutrient and raw material, is of great importance to the holistic agriculturalist. Its use ensures that the pitfalls, so easy to make when using by-products, are avoided. The pitfalls are usually associated with wrong energy/protein balance, or deficiency of an essential nutrient – usually an amino acid – lead to disastrous performance by the animals. The appropriate agriculturalist might be forgiven for renaming "least-cost formulation" "best-use formulation" because by using a computer the best utilization of products is made.

The computer must be provided with the cost and nutrient make-up of all the possible raw materials which may be available, and additionally the specification in nutrient and raw material terms of the diet to be formulated. Often restrictions need to be placed on the use of certain ingredients which are available; for instance an ingredient may be in short supply and therefore has to be used sparingly, or the ingredient may be unpalatable if used at too high a percentage incorporation. Such an ingredient is maize gluten meal, which poultry find unpalatable if fed at levels above 5% of the diet.

Once given the specification the computer will produce within seconds a formula and analysis for the diet.

Several precautions have to be taken when formulating diets based on a high proportion of by-products. Firstly the vitamin and trace element make-up of the ingredients may not be known, and to overcome this difficulty it is usual to include as a compulsory ingredient a vitamin/trace element mineral supplement at 1.25% of the diet. Supplements made by a reputable company will meet the vitamin and trace element requirement for the class of livestock for which they are intended. All vitamins have natural sources, and these can be used instead of dry supplements. Usual sources are fish liver oil for vitamins A and D_3; dried yeast for the B group; green foods and vegetable oils for vitamins

C and E. Seaweed meal is often used as a source of macro- and micro-nutrients for livestock. It is an excellent product and can be used without harm for all livestock.

Many by-products may be deficient in lysine, which is important in pig diets, and methionine in poultry diets. Synthetic lysine and methionine are available, and can be used to advantage, but the routine feeding of these synthetic amino acids, like the use of urea in crop husbandry, should not be countenanced as they give rise to an unsustainable agriculture. Far better to set a specification for the amino acid in the diet and allow the computer to formulate to the standard. The practice of using urea as a degradable protein source for ruminants is never acceptable. Under holistic farming systems where silage is made the rumen-degraded protein is never limiting, and the feeding of urea (which only provides rumen-degraded protein) will cause unbalanced bacterial growth in the rumen and possible excessively high blood urea levels, with consequential metabolic upsets.

8.1   Diets for Monogastrics

Baby Chicks

Table 8.1 is a representation of a computer printout for a baby chick diet. The printout shows the nutritional specification which was demanded, as well as constraints on the usage of various raw materials which were available. It will be noted that the computer, besides producing the formula and analysis of the diet, also gives additional information of value to the formulator.

Cost Ranging

This part of the table shows the upper and lower limits for the cost of raw materials before a formulation change occurs. For instance if biscuit (high oil) costs more than £107.65 per tonne then the inclusion rate will drop to 11.539%.

Rejected

In this section the computer evaluates the price at which the rejected materials would have to be sold for them to be used in the formula. For instance, barley which costs £118.00 would be used in the diet at 7.032% if the price dropped to £109.78.

Sensitivity

In this final section the computer calculates the cost, or saving, which can be made by altering the specification restraints. More often than not constraints

TABLE 8.1

Diet: Chick
Optimal Cost £129.05

Formula	%	Min.	Max.	Cost
Wheat	37.447	—	—	120.00
Soya ext.	14.231	—	—	163.00
Peas	4.464	—	—	130.00
Sunflower ext.	3.301	—	—	108.00
Fish Meal 8–10% oil	2.500	2.50	—	295.00
Biscuit hi. oil	13.332	—	25.00	100.00
Stale bread	3.476	—	—	80.00
Wheatfeed 6% fib	15.000	—	15.00	105.00
Maize	5.000	5.00	—	190.00
Vit/min supp.	1.250	1.25	1.25	225.00

Analysis

Met. energy	11.95	11.81	11.95	—
Protein	18.051	17.50	—	—
Methionine	0.370	0.35	—	—
Methionine + cyst.	0.670	0.67	—	—
Fibre	4.000	—	4.00	—
Linoleic acid	0.831	0.80	—	—
Sodium	0.191	—	0.191	—
Oil	4.272	—	—	—
Calcium	0.614	—	—	—
Phosphorus	0.518	—	—	—

Cost Ranging	Lower limit	% inc.	Upper limit	% inc.
Wheat	112.91	39.38	121.08	32.801
Soya ext.	160.16	15.997	169.77	10.586
Peas	119.03	6.712	132.29	2.274
Sunflower ext.	74.81	4.044	159.89	2.128
Biscuit hi. oil	95.17	13.332	107.65	11.539
Fish meal 8–10%	177.66	7.291	—	—
Stale bread	60.17	4.167	92.43	3.072
Wheatfeed	—	—	118.32	0.000
Vit/min supp.	—	—	—	—

Rejected	Cost	Value	% inc.
Barley	118.00	109.78	7.032
Meat and bone meal	178.00	125.66	4.718

Sensitivity	At	Saving	At	Costing
Fibre	4.52	1.46	3.74	0.72
Metabolizable energy	12.63	0.69	11.76	0.14

do cost the farmer money, but generally it is not possible to alter the specifications as the nutrition of the animal would suffer.

The reader will observe that the printout of the formula is mathematically exact, and it would not be possible to mix this formula on the farm. The formulator will therefore "round" the formula to that which is practical. It is usually impossible to round to less than 1.25% which, for food bought in sacks, is one half of a 25 kg sack per tonne. The chick formula given in Table 8.1 when rounded therefore becomes as shown in Table 8.2.

TABLE 8.2

Ingredient	Percentage	kg per tonne	cwt per ton
Wheat	36.75	367.5	$7\frac{1}{3}$
Soya ext.	15.00	150.0	3
Peas	5.00	50.0	1
Sunflower ext.	2.50	25.0	$\frac{1}{2}$
Fish meal 8–10%	2.50	25.0	$\frac{1}{2}$
Biscuit meal	13.25	132.5	$2\frac{2}{3}$
Stale bread	3.75	37.5	$\frac{3}{4}$
Wheatfeed	15.00	150.0	3
Maize	5.00	50.0	1
Vitamin/minerals	1.25	12.50	$\frac{1}{4}$
	100.00%	1000.0 kg	20 cwt

Laying Hen Diets

Table 8.3 is a copy of a computer printout for a laying hen diet. The formula given by the computer is always for a complete diet, but the farmer will often want to feed different constituents at different times of the day. For instance with laying hens some wheat will be fed as a "scratch feed" and the calcium required (formulated as limestone flour), will generally be given as oyster shell on an *ad lib.* basis.

As an example of the alterations to be made to the computer printout the

TABLE 8.3

Diet: Layer
Optimal Cost £119.72

Formula	%	Min.	Max.	Cost
Wheat	32.001	25.0	—	120.00
Soya bean meal	14.175	—	—	163.00
Sunflower ext.	12.961	—	—	108.00
Biscuit hi. oil	23.063	—	—	120.00
Maize gluten feed	5.000	—	5.00	110.00
Limestone flour	9.494	—	—	25.00
Fish meal white	1.250	1.25	—	275.00
Fat/oil	0.806	—	—	300.00
Vit/min supp.	1.250	1.25	1.25	225.00

Nutrient	Analysis	Min.	Max.
Protein	17.50	17.50	—
Metabolizable energy (MJ/kg)	11.72	11.72	—
Linoleic acid	1.30	1.30	—
Fibre	5.00	—	5.00
Methionine	0.43	0.35	—
Methionine + Cystine	0.73	0.66	—
Calcium	4.00	3.20	4.00
Phosphorus	0.39	—	0.50
Oil	6.20	—	—
Ash	12.91	—	—
Sodium	0.12	—	—

TABLE 8.4

Ingredient	Percentage included
Wheat	31.50
Sunflower meal	12.50
Biscuit meal (high oil)	22.50
Maize gluten feed	5.00
Limestone flour	10.00
White fish meal	1.25
Fat/oil	1.00
Vitamin/minerals	1.25
Soya bean meal	15.00
	100.00

following rounded first stage layers diet are shown in Table 8.4. The addition of oil to this diet will reduce the dustiness of the ration, besides being a very rich source of linoleic acid – an essential fatty acid for hens.

In this example the farmer has decided to feed oyster shell instead of the limestone flour, thus taking out 10% of the above contents. He further wishes to feed 25 g per day of wheat as a scratch feed, removing a further 20% of the contents (25 g is about 20% of a laying hen's daily food intake).

The formula for the remaining diet to be fed as mash is now 70% of the original whole, and must therefore be proportionally adjusted by multiplying (in this case) the percentage inclusion of each ingredient by 100 and dividing by 70. The formula for the mash therefore becomes:

11.5% Wheat (31.5–20% fed as scratch) $\times$ 100 $\div$ 70 = 16.43
15.0% Soya bean meal $\quad\quad\quad\quad\quad$ $\times$ 100 $\div$ 70 = 21.43
12.5% Sunflower meal $\quad\quad\quad\quad\quad$ $\times$ 100 $\div$ 70 = 17.86
22.5% Biscuit meal (high oil) $\quad\quad$ $\times$ 100 $\div$ 70 = 32.14
5.0% Maize gluten meal $\quad\quad\quad\quad$ $\times$ 100 $\div$ 70 = 7.15
1.25% White fish meal $\quad\quad\quad\quad$ $\times$ 100 $\div$ 70 = 1.78
1.00% Fat/oil $\quad\quad\quad\quad\quad\quad\quad$ $\times$ 100 $\div$ 70 = 1.43
1.25% Vitamin/mineral supplement $\times$ 100 $\div$ 70 = 1.78
70% of original diet $\quad\quad\quad\quad\quad$ $\times$ 100 $\div$ 70, 100.00% of new diet.

or when rounded to a practical formulation for every tonne:

Wheat	157.5 kg
Soya bean meal	212.5 kg
Sunflower meal	175.0 kg
Biscuit	325.0 kg
Maize gluten	75.0 kg
White fish meal	20.0 kg
Fat/oil	15.0 kg
Vit/min supp	20.0 kg
	1000.0 kg

Weaner Pig Diets

Table 8.5 is an example of a printout for weaner pigs which utilizes skim milk. In the table the skim milk is shown in the "dry form", so once again there has to be adjustment to the formulation to give the exact quantities to be mixed with the liquid skim.

TABLE 8.5

Diet: Weaner
Optimal cost: £130.93

Formula	%	Min.	Max.	Cost
Wheat	20.946	—	—	120.00
Soya bean meal	14.104	—	—	163.00
Peas	5.000	—	5.00	130.00
Sunflower ext.	13.422	—	—	108.00
Milk skim dried	15.000	—	15.0	152.00
Biscuit high oil	25.000	—	25.0	120.00
Cassava/manioc	5.000	—	5.0	105.00
Tallow	0.278	—	3.0	340.00
Vitamin/mineral	1.250	1.25	1.25	225.00

Nutrient	Analysis	Min.	Max.
Fibre	5.000	—	5.0
TDN	77.000	77.0	—
Available lysine	1.000	1.0	—
Oil	5.722	—	—
Protein	20.555	—	—
Ash	4.868	—	—
Calcium	0.773	0.65	—
Phosphorus	0.479	—	—
Sodium	0.105	—	—
Total lysine	1.184	—	—

The calculations which have to be made are as follows. The computer has indicated that the ration must contain 15% of dried skim, i.e. 150 kg of dried powder per tonne of feed, or because skim milk is 10% dry matter, 1500 litres of liquid skim. This quantity of liquid skim would be mixed with 850 kg of meal made to the proportionally adjusted formula (see the previous layers mash example).

The "dry" ration to feed with the skim milk then becomes:

272.5 kg of ground wheat
162.5 kg of soya bean meal
 50.0 kg of ground peas
150.0 kg of sunflower meal
300.0 kg of biscuit meal
 50.0 kg of cassava or manioc
 15.0 kg of vitamin/mineral supplement

Finishing Pig Diets

Table 8.6 is again a computer printout for a finishing pig diet. The diet contains 49.63% of dried household swill. Swill has an average dry matter content of 27% and therefore 496.3 kg of "dry swill" is approximately the same as 2 tonnes of cooked swill as collected. Therefore by proportion, to this 2 tonnes of cooked swill must be added 265 kg of soya bean meal, 225 kg of extracted rice bran, and 12.5 kg of a vitamin mineral supplement. Salt would not be added to diets where whey or swill are fed. In this example, because the swill is of very high energy value and is low in protein, it has to be balanced by the addition of a protein source (soya bean meal) and a "filler" (extracted rice bran). Oatfeed, a by-product of the porridge oat industry, would be another suitable filler to use. Additionally when swill is fed the farmer must ensure that the diet does not have too high a moisture content. This case shows a dry matter content well above the critical 22.5–25.00%.

TABLE 8.6

Diet: Finisher
Optimal cost: £90.41

Formula	%	Min.	Max.	Cost
Soya bean meal	26.655	—	—	163.00
Vitamin/mineral	1.250	1.25	1.25	225.00
Raw swill	49.630	—	—	55.00
Ext. rice bran	22.465	—	—	75.00

Nutrient	Analysis	Min.	Max.
TDN	76.000	75.00	76.00
Available lysine	0.850	0.850	—
Oil	7.058	—	—
Protein	23.782	—	—
Fibre	7.073	—	—
Ash	4.239	—	—
Calcium	0.674	0.65	—
Phosphorus	0.652	—	—
Sodium	0.125	—	—
Total lysine	1.160	—	—

The quantity of food fed to the finishing pig varies according to the genetic potential of the animal, as well as the strictness of the grading standards demanded. In the U.K. a finishing pig would receive about 2 kg of dry meal per day. Therefore the above-mentioned mixture would feed about 500 finishing pigs for 1 day.

8.2   Diets for Ruminants

Cattle and Sheep

Designing diets for cattle and sheep is more complicated than for the monogastric hen and pig. Allowances have to be made for degradation of food by the rumen micro-organisms which may reduce the level of some of the amino acids fed, but because some of the organisms die and are carried through into the digestive tract proper they become a source of dietary protein for the animal. It is now standard practice to design all ruminant diets on a dry matter basis, so obviating the difficulty of making allowance for the varying dry matter contents of all the different foods which will be consumed. To complicate the issue further, the ruminant nutrient requirements vary according to the digestibility of the whole diet which is to be fed. The percentage of gross energy of the diet which is metabolizable is normally expressed as a decimal and is designated by the letter q. The values of q for different feeds are given in the analysis tables. However for roughages they have to be calculated. Luckily gross energies are remarkably constant at about 18.4 MJ/kg of dry matter for hay, and 19.8 MJ/kg for silage and grass. The metabolizable energy for any roughage can be determined from the chemical analysis using one of the following formulae:

Hay metabolizable energy $= 14.3 + 0.17 \times cp - 0.19 \times dm$
Silage metabolizable energy $= 10.9 + 0.21 \times cp - 0.047\ madf - 0.06\ dm$
Maize silage $= 12.3 - 0.12\ madf$

where cp is the crude protein; madf is the modified acid detergent fibre; dm the dry matter, all expressed as percentages.

As has already been stated, some of the protein consumed by the ruminant is degraded in the rumen by the micro-organisms present in the rumen. In diet formulation it is important to know the degradability of the various proteins and again analysis tables give this information for common feedstuffs. Agricultural scientists have agreed that for foods where degradability has not yet been determined by experiment, the following degradabilities will be used:

Barley, wheat, oats, and oil seed cakes   70% degradable
Maize, lupins, fish meals   50% degradable
Milo, meat and bone meals   30% degradable

Degradability of protein in roughages has to be determined and the best estimate of the value is given by the formula:

$$\frac{0.9\ (\text{Crude protein} - 0.1\ \text{Modified acid detergent fibre})}{\text{Crude protein \%}}$$

Finally all nutrients must be contained within the dry matter appetite of the animal (tables for appetite are given in the Appendix), and so that the rumen functions correctly at least 40% of the daily dry matter intake must be in the form of roughage foods.

Once all the necessary factors have been determined then diet formulation is comparatively simple.

Consider an Ayrshire cow giving a daily yield of 18 litres of milk. The foods available are silage, barley, wheat, and tic beans. The silage analysis is:

16% crude protein
36% modified acid detergent fibre
0.35% calcium
0.75% phosphorus
18.00% dry matter

From these data, using the formulae already provided, the metabolizable energy is calculated to be 11.5 MJ/kg of dry matter and the degradability of the silage is calculated to be 70%, therefore of the 16% protein in the silage 11.2% will be degraded in the rumen and the remaining 4.8% will not be degraded and is available for digestion in the lower reaches of the digestive tract.

The nutrient requirements per day (obtained from the appendix tables) for an Ayrshire cow giving 18 litres of milk on a diet where $q = 0.6$ are as follows:

Metabolizable energy (ME) 163 MJ
Rumen-degraded protein (RDP) 1300 g
Undegraded protein (UP) 325 g
Crude protein 1625 g
Calcium (Cal) 74 g
Phosphorus (Phos.) 60 g

The nutrients are to be contained in 16 kg of dry matter (the cow's daily dry matter appetite) of which 40% (6.4 kg) must be obtained from roughage (Table 8.7).

TABLE 8.7

	Dry matter (kg)	ME (MJ)	RDP (g)	UDP (g)	Cal. (g)	Phos. (g)
Total	16.0	163	1300	325	74	60
Silage provides	6.4	75	717	307	22	48
Difference to be provided by dry food	9.6	90	583	18	52	12

Therefore 1 kg of 'dry' food must provide:

9.4 MJ of energy (90 ÷ 9.6)
61 g of rumen-degraded protein (583 ÷ 9.6)
2 g of undegraded protein (18 ÷ 9.6)
6 g of calcium (52 ÷ 9.6)
1.25 g of phosphorus (12 ÷ 9.6)

and these nutrients would be adequately met by feeding rolled barley with the addition of a vitamin trace mineral supplement which was carried on limestone flour.

When designing ruminant diets it is of great importance to try to balance the diet so that there is no excess of rumen-degraded protein. Excess of this nutrient can cause the animal's blood urea level to rise, and this can be the commencement of many of the metabolic upsets seen in cattle fed conventional high-protein cakes. In most ruminant diets, in the U.K. at least, it is energy and major minerals which will be limiting.

With good silage as roughage the only dry food required in the example shown would be barley and vitamin mineral supplement. If, however, the only roughage available were poor-quality hay then the supplementary feed would need to include at least tic beans and perhaps fish meal to meet the undegraded protein needs. It will therefore be readily seen that it is of the utmost importance that high-quality, high dry matter silage, as discussed in the chapter on grassland, is obtained. This drier, higher-quality silage not only saves the cost of expensive purchased proteinaceous cake, but also makes the livestock enterprise more sustainable within the farm's own resources.

As with the non-ruminants, the metabolizable energy of foods other than roughages may need to be calculated, and the following equation gives a good estimate of the metabolizable energy per kilogram of dry matter.

$$MER = 0.12 \, cp + 0.31 \, oil + 0.05 \, cf + 0.14 \, nfe$$

where cp = crude protein percentage in the dry matter of the food;
   oil = percentage of oil in the dry matter of the food;
    cf = percentage crude fibre in the dry matter of the food;
  nfe = percentage of nitrogen free extract in the dry matter of the food.

Goats

Unfortunately the nutritionists have not given as much attention to the feeding of goats as they have to other farm animals, and the only dietary standards for milking and mohair-producing goats have only recently been published by the American National Research Council. The standards they suggest are included in Appendix II.

Although the American figures only give values for a total crude protein requirement for all classes of goats, it seems reasonable that as the goat is a ruminant it will have a demand for undegradable protein, and as a precaution it may be that all diets for heavily milking goats should contain 1.25% of animal protein.

Again the same authorities do not suggest what percentage of the daily diet should be in the form of roughage, and it would appear sensible to utilize the goat's browsing ability to ensure that at least 50% of the daily intake is in the form of roughage.

Even less is known about the amino acid requirements of the goat, and in all probability milking goats fed protein at the levels suggested will not have their milk production limited by any amino acid deficiency. However, it appears possible that, for goats kept for mohair production, the quality and quantity of mohair may be dependent on adequate supplies of methionine, and again the inclusion of a small quantity of animal protein in the concentrate part of the diet would seem to be indicated.

Chapter 9

Some Other Considerations

ANY new agricultural system, before being implemented on a large scale, must be able to demonstrate that it can produce food in sufficient quantity to feed the world population and that the production of this food can be at a cost, in terms of human dignity and finite raw materials as well as finance, comparable or better than the conventional system it is intended to replace. Additionally the food produced by the system should be of equal or better nutritional quality than that produced by the present farming methods.

Holistic agriculture has therefore to be investigated with particular reference to all these points, and the purpose of this chapter is to show how this appropriate system answers the criteria posed. Once the method has been shown to answer the criteria positively, then the practical methods can be worked out by the farmer on his particular farm.

9.1 Yield

The subject of yield should be divided into considerations of yield of food produced by plants, and that produced by the livestock enterprises. Since 1972 several investigations, on a short time basis, have been made into the yields and profitability of various crops.

A study carried out by the Soil Association[1] in conjunction with the Department of Land Economy at Cambridge University investigated the yields of winter wheat and spring barley grown under ecological methods, and these results were compared with the Cambridge University surveys carried out in those years. The results are given in Table 9.1. The study showed that the yields of the two crops varied enormously, from farm to farm under both farming methods, but the total variation under the two systems did not differ significantly from the variation normally seen. Indications were that the crops with high nutrient requirements did not yield as heavily under the organic systems, but barley did yield as well, if not better when grown ecologically. Lockeretz et al.,[2] in an experiment conducted in 1974 and 1975, demonstrated that in the mid-west of America crops with high nutrient demand, such as maize, showed an overall reduction in the two years of about 13% when grown without agrochemicals compared with those crops grown conventionally

143

TABLE 9.1 *Gross margin analysis for winter wheat 1972 and 1973*

| | Organic production | | | | | | | | Average Cambridge Survey | |
| | Case Study 1 | | Case Study 2 | | Case Study 3 | | Average | | | |
	1972	1973	1972	1973	1972	1973	1972	1973	1972	1973
Yield (cwt per acre)	33.40	33.70	36.20	38.30	32.20	31.90	33.90	34.60	38.00	34.80
Gross output (£ per acre)	71.48	136.30	74.90	155.90	63.70	118.20	70.02	136.80	62.90	69.60
Total variable costs	3.61	8.00	10.80	13.40	5.20	6.31	6.53	9.23	12.80	15.50
Gross margin per acre	67.87	128.30	64.10	142.50	58.50	111.89	63.49	127.57	50.10	54.10

TABLE 9.2 *Mean economic performance, organic and conventional samples. US$/ha. Interquartile range in brackets. After Lockeretz 1984*

| Year | Value of crops produced | | Operating expenses | | Net returns | |
	Organic	Conventional	Organic	Conventional	Organic	Conventional
1974	393 (348–469)	426 (346–501)	69 (49–84)	113 (96–131)	324 (264–380)	314 (249–383)
1975	417 (326–479)	478 (422–479)	84 (72–101)	133 (104–148)	333 (269–385)	346 (294–353)
1976	427 (371–534)	482 (407–524)	91 (70–109)	150 (126–164)	336 (282–427)	333 (270–377)
1977*	384	407	95	129	289	278
1978†	440	527	107	143	333	384

*23 farms not the same as 1974–76, compared with regional averages.
†19 farms, also not the same as 1974–76.

TABLE 9.2(a) *Yield per acre*

	Organic		Conventional		Percentage organic less than conventional
	1975	1974	1975	1974	
Corn (bu)	74.0	74.0	94.0	76.0	13
Soybeans (bu)	35.0	32.0	38.0	29.0	0
Wheat (bu)	28.0	28.0	38.0	29.0	16
Oats (bu)	58.0	56.0	60.0	60.0	5
Hay (tons)	4.5	5.0	3.9	3.4	−30

(Table 9.2a). However leguminous crops grown organically had the same yield as conventional ones during the experimental period. Hay production was 30% better under the ecological system. Lockeretz made the point that in 1974 differences in yield for any crop were not significant, whilst in 1975, an excellent growing year, more response was obtained from the fertilizer used in the high nutrient demanding crop.

More recent studies have been carried out at the Pye Research Centre farm in Haughley, Suffolk. They show quite clearly that winter wheat yields are lower, whereas spring barley and oat yields are greater than the chemically grown crops in similar farming areas. The results of the investigation carried out by the Land Economics Division of Cambridge University are given in Table 9.3.[3] Under the present regime of financial structure the high-nutrient intake crops, such as wheat and maize, are the high-profit crops in any particular country, and tend therefore to be grown by all types of farmers, whereas it is possible that if ecological farmers grew crops more ecologically acceptable to their particular land and climate, then their yields would be the same or better. This is certainly true when the results from the Haughley research farms are considered. The research farms have very good wheat land, and the yields for the ecological unit are excellent and not significantly different from those grown chemically on the same soil types in the same climatological area.

TABLE 9.3

	Organic section		Chemical section	
	Yield (ton/ha)	Gross margin (£/ha)	Yield (ton/ha)	Gross margin (£/ha)
Winter wheat	5.29	593.00	5.70	449.40
Spring barley	5.04	500.00	4.24	322.70
Oats	5.17	499.00	4.64	353.80

9.2 Profitability

It is important that profitability to the grower is considered. Lockeretz's (1975) study did not investigate profitability, but the other two investigations already discussed did consider this aspect. At the present time "organic" grain commands a high premium, but these two studies show that, when corrected for normal cereal prices, gross outputs for wheat grown organically will be lower than those attained by the chemical farmer. But because the total variable costs on ecological systems are lower than on chemical systems, the gross margins are generally in favour of the organic grower. The figures obtained from the more recent Haughley study (Table 9.3) show this increase. The results have an implication for total farm profitability in that the total amount of money required to finance the farming operation will be lower, and therefore the interest on working capital is lower and final profitability of the cereal enterprise should be higher. The proviso must be made (and this is discussed later) that there appears to be an increased labour charge with ecological methods. The other interesting point which is highly relevant to the financial debate is that owing to the escalation of fertilizer prices since 1974 (due to the increase in energy costs) the total variable costs for the chemical farmers have increased eight times more than those of the organic farmer. This single fact is one of the reasons why farmers and economists in general are becoming more interested in the possibilities of ecological farming.

Lockeretz *et al.*, in a study published in 1984,[4] updated and extended previous work. In this later work they showed the value of the crop, operating expenses and net return for organic and conventional farms, and perhaps more importantly gave, where possible, the inter-quartile range for these parameters. The results are shown in Table 9.2. The reader should particularly note the great variability in each parameter for both types of farming.

With regard to livestock enterprises, the survey which was conducted by the Soil Association in conjunction with Cambridge University is summarized in Table 9.4. This once again shows the tremendous variation in performance and profitability from farm to farm. What is certain is that the best farms on either system were equal to each other in yield, gross output, and gross margin. Oelhaf's[5] analysis of the Guernsey herds, held at the Haughley research farms, shows that the outputs of milk per acre from the organic and chemical sections of the farm were not significantly different; but towards the yield per cow and the amount of food fed per gallon on the organic section was significantly better than those obtained on the chemical section. There would be general agreement nowadays that milk production under ecological systems is equal to or better in yield and profitability to that obtained under conventional methods. This result may in some way be due to the generally better livestock management seen in the smaller ecological farms, or to the higher dry matter content of the home-grown food.

The final aspect of animal husbandry which must be considered is that of the beef and sheep enterprises. So far the author is not aware of any statistical

TABLE 9.4 *Gross margin analysis, £ per cow, 1973/4*

	Organic Farms						Average conventional farms (milk costs enquiry)
	Farm 1 Ayrshire	Farm 2 Jersey	Farm 3 Friesian	Farm 4 Ayrshire	Farm 5 Ayrshire	Average all breeds	
Milk sales	220.30	208.93	214.20	186.02	153.20	196.53	181.21
Calf sales	17.70	12.86	32.60	8.73	13.05	16.99	24.19
Total sales	238.00	221.79	246.80	194.75	166.25	213.52	205.40
Variable costs							
Concentrates purchased and home grown	68.50	54.70	72.30	62.49	41.60	59.92	47.87
Margin over concentrates	169.50	167.09	174.50	132.26	124.65	153.60	157.53
Forage variable	4.80	5.26	3.01	5.29	5.93	4.86	10.12
Bulk feed purchased	7.30	—	—	10.54	—	3.57	2.30
Misc. vet, A.I., sundries	9.60	6.54	12.40	6.41	6.30	6.88	8.17
Gross margin/cow	147.80	155.29	159.09	110.02	112.42	136.92	136.97
Gross margin/forage acre	140.86	146.50	98.87	68.90	53.27	101.68	97.12

studies which have been made on these livestock enterprises, but the 1972/3 Soil Association survey showed no statistical difference in the forage acres per cow under ecological or chemical systems, and it would therefore not be unreasonable to extrapolate that the stocking rates for beef and sheep would be comparable under the two systems, but again as the total variable costs will be lower ecologically, then the gross margins will once again favour the ecological system.

Mr Murphy of Cambridge University, who supervised the Soil Association survey, summed up his report by stating:

> The study has shown that it is possible for organic farmers to be as efficient as conventional farms in terms of yield per acre, or per cow. That is to say that individual enterprises in cereals or dairying achieve levels of physical and economic performance which compare with published standards.

In 1981 Vine and Bateman[6] in a report of a study of organic farms in the U.K. for just one farming year, stated in their summary:

> Although the majority of organic farmers had lower financial returns, it should be noted: firstly that the variety of systems and levels of management were such that *some* organic farmers out performed the conventional average.

As recently as November 1986 Bateman and Lampkin[7] at the University College of Wales Agricultural Society conference said:

> In spite of the paucity of data organic farmers appear to achieve financial results which are similar to or slightly lower than conventional farms. Lower yields tend to be compensated for by a combination of reduced input costs and premium prices leading in many cases to higher gross margins.

All the evidence from agricultural economists supports the self-evident truth that there are good and bad ecological farmers, and good and bad chemical farmers, but what cannot be in dispute is that under good management, results in terms of yield and profit by the former are as good as those attained by the best of the latter. Therefore it can be concluded that holistic farming, properly practised, will produce yields comparable to conventional methods, and the gross margins of production are in general better than those attained by conventional means.

9.3 Food Quality

Many experiments have been conducted over the years, but there is still not complete statistical evidence on the effect of the types of food in respect of human nutrition. There is, however considerable experimental work to show

TABLE 9.5 *Some changes which occur when composted manure is used instead of chemical fertilizer (results for chemical fertilizer treatment = 100%)*

Parameter	Composted manure	Percentage difference
Total yield	76	−24
Yield of dry matter	123	+23
Nutritionally valuable		
True protein	118	+18
Vitamin C	128	+28
Total sugar	119	+19
Methionine	123	+23
Minerals:		
Potassium	118	+18
Calcium	110	+10
Phosphorus	113	+13
Iron	177	+77
Nutritionally undesirable		
Nitrate	7	−93
Simple nitrogen compounds	58	−42
Sodium	88	−12

that food produced ecologically has a greater percentage of the nutritionally desirable constituents and a far lower content of the undesirable nutritional factors. The most comprehensive survey was undertaken by Schupham,[8] who in a 12-year experiment into food content produced the results given in Table 9.5.

Table 9.6 shows that storage losses in vegetables always appear to be less when the crop has been grown ecologically.

It would therefore appear that, although there is no proof, there is certainly no evidence to show that the food produced by chemical methods is superior to that produced ecologically. Indeed the converse is apparently true.

TABLE 9.6 *Storage losses in various products of different origin by percentage (after Vogtmann)[9]*

Product	Conventional	Ecological
Potatoes	24.5	16.3
Potatoes	30.2	12.5
Various vegetables (average)	46.2	30.0
Carrots	45.2	34.5
Turnips	50.5	34.8
Beetroot	59.8	30.4

9.4   Energy Utilization

The animal kingdom is not able to utilize the sun's energy directly, but has to rely on the energy of carbohydrate which the plants have synthesized from carbon dioxide and water, using the sun's energy in the photosynthesis process.

Agriculture, and for that matter horticulture, is concerned with the organized cultivation of plants to produce carbohydrate, some of which will be directly consumed by man, whilst some, not of direct use to man as food, will be converted by animals into a form which can be used by man. Obviously this conversion through the animal cannot be as efficient as direct plant consumption, but is justified in that the animal should use the carbohydrate which can only be produced on land where crops for human need cannot be grown, or those grown on land which has to be rested in the interests of good crop husbandry – such crops are muka and grassland. Additionally the animal should receive back as food the by-products made available by the conversion of the plants, sold direct for human food, into an edible form (for instance oat feed or wheatfeed by-products of the milling industry). There will also be available for animal consumption food wasted in home preparation and that which is left on the plates when the meal is concluded. This home waste amounts to nearly 4% of the food brought into the home.[10]

Blaxter has investigated the efficiency of modern chemical British agriculture from an energy standpoint and Tables 9.7–9.9 summarize his findings.

TABLE 9.7

Energy inputs	$MJ \times 10^9$ per year	Percentage of total
Fuel		
Coal	4.3	1.26
Petroleum products	82.8	24.33
Electricity	57.4	16.86
Total fuel	144.5	42.45
Fertilizers		
Nitrogenous	94.6	27.79
Phosphatic	8.4	2.47
Potassic	3.9	1.14
Lime	21.2	6.73
Total fertilizers and lime	128.1	37.63
Agrochemicals	1.2	0.35
Machinery		
Depreciation		
Repairs	48.8	14.34
Proportion of contract use		
Process of feedstuffs	2.1	0.62
Transport		
To the farm (fuel fertilizers, lime, feed, etc.)	3.5	1.03
From the farm (crops and stock)	12.2	3.58
Total energy consumption	340.4	100.00

Data from Blaxter.[11]

As can be seen, Blaxter estimates that U.K. agriculture only produces about 58% of the food energy requirements of the population, and more depressingly that which it does produce, uses more energy than the food it sells, let alone the food eventually consumed.

Closer examination of the tables reveals that the total energy required for the production of fertilizers just about equates with the total energy deficiency. If, as was discussed in an earlier chapter, all the human and some of the household garbage were returned to the land there would be no loss of nutrient and manufactured fertilizer would virtually cease to be required. The energy of transportation of the nutrient from the urban areas would be slightly increased as the composted material would be less concentrated than present-day chemical fertilizers.

It will also be noted from the tables that about 80% of the energy produced in the form of cereals remains on farms as feed or bedding for animals. Obviously any steps which can be taken to reduce the dependence of the country's livestock on cereals helps make agriculture more sustainable. For instance swill, surplus to the needs of the compost manufacturer, used for pig feed could reduce this dependence by at least 0.7 million tonnes of grain, which is equivalent to the production of 10% of the land allocated to barley growing.

TABLE 9.8

	Plant products			
	Energy yield $(MJ \times 10^9/year)$			Percentage of production retained on farm
Energy outputs	Produced	Sales off farm	Consumed	
Crop				
Wheat	81.8	36.0	25.9	56
Barley	152.0	27.2	9.5	72
Oats	20.5	2.0	1.3	90
Other grains	6.6	—	—	100
Straw	80.0	—	—	100
Cereal total	340.9	65.2	36.7	81
Potatoes	23.8	18.2	16.4	23.5
Sugar beet	25.7	25.7	4.4	—
Turnips	5.9	—	—	100
Mangolds	0.7	—	—	100
Fruit	1.6	1.6	1.5	—
Vegetables	8.1	8.1	5.7	—
Total other arable crops	65.8	53.6	28.0	18.5
Grass	390.0	—	—	100
Rough grazing	176.0	—	—	100
Hay/silage	143.5	—	—	100
Total grass and grass products	709.5	—	—	100
Grand total	1116.2	118.8	64.7	89

(*continued*)

TABLE 9.8 *continued*

	Energy Yield − MJ × 10⁹/year	
	Total output	Consumed
Animal products		
Poultry		
Broilers	6.0	3.9
Hens	1.2	0.8
Eggs	5.3	5.2
Total poultry output	12.5	9.9
Pigs		
Pork	2.6	1.8
Cutters	3.2	2.2
Bacon	7.6	5.4
Heavy pigs	2.4	1.8
Sow and boar meat	0.8	0.6
Total pig output	16.6	11.8
Sheep		
Lambs	4.1	—
Cull ewes	1.0	—
Rams	0.1	—
Total sheep output	5.2	2.9
Cattle		
Calves	0.1	—
Medium weight bullocks	13.2 ⎫	
Heavy weight bullocks	6.9 ⎭	15.2
Milk	39.5	33.2
Total cattle output	59.7	48.4
Total all livestock	94.0	73.0

Data from Blaxter.[11]

TABLE 9.9(a)   *Energy balance sheet for British agriculture*

Inputs (MJ × 10⁹/year)		Outputs (MJ × 10⁹/year)	
Total energy inputs from		Farm gate output, crops	119
Table 9.7	340	Farm gate output, animal	94
		Total farm output	213
		Annual energy deficiency	127
	340		340

TABLE 9.9(b)   *Energy balance sheet for U.K. population*

MJ × 10⁹/year		MJ × 10⁹/year	
Energy need of U.K.		Energy from crops eaten	65
population	241	Energy from livestock	
		products consumed	73
		Deficiency imported food	93
	241		241

Data from Blaxter.[11]

9.5 Finite Resources

With regard to the use of finite resources the position is no less depressing. The major plant nutrients phosphorus and potassium are at present mined, and any residues which occur in faeces or urine are by and large discharged into the sea via sewerage systems. It has been estimated that to convert the phosphorus lost to the sea back to mineable quantities will need a time span of about 10 million years. In respect of potassium, U.K. supplies exploited since World War II are no longer commercially viable, and practically all U.K. potash needs have to be imported from other countries. Obviously the raw materials needed to support the present chemical agriculture are becoming less available and this, together with the deficit energy balance, makes present agricultural methods look less than secure – chemical agriculture is non-sustainable.

Consideration must therefore be given to ascertain whether or not ecological agriculture is more secure. Work carried out at the Dutch Research Station at Nagele in the Netherlands has investigated, amongst other subjects, the loss of nutrients from the soil in an intensive agricultural holding. Their findings are given in Table 9.10.

TABLE 9.10 *Annual nutrient losses in kg per year from an intensive 22 ha farm*[12]

	Nitrogen	Phosphorus	Potassium
80 tonne milk; 3.35% protein	430	180	180
1.5 tonne beef	50	20	30
7 ha grain at 4.5 t/ha	450	100	150
2 ha potatoes at 25 t/ha	150	40	250
Leaching losses on 22 ha	1320	0	990
NH$_3$ volatilization NO$_3$ denitrification } on 22 ha	1540?	0	0
Gross loss per ha per year	179	15.5	73

Dealing with the nutrient losses in some detail it will be noted that there is a loss of 179 kg of nitrogen per hectare. This can easily be restored, not only by the use of legumes (the F.A.O.[13] show that legumes can produce over 500 kg of nitrogen per hectare), but by reducing the massive leaching losses. The losses would be substantially reduced by the use of green manures. In all probability an ecological farm using green manure techniques and legumes is in positive nitrogen balance, when the whole farm is considered on a rotational basis. The indicated loss of 15 kg of phosphorus must be considered. Work at Haughley research farms has shown that the quantity of available phosphorus is directly proportional to the organic carbon content of the soil, and this factor is correlated with the organic matter in the soil. Experiments at Haughley show that as much as 50 ppm of phosphorus become available under ecological systems – this being due to the biological activity in the soil (mainly mycor-

TABLE 9.11

Farming type	Organic carbon (%)	Exchangeable potassium (m.e./100 g soil)	Available phosphorus (ppm)	Cation exchange capacity (m.e./100 g)
Permanent grass	4.62	0.79	58.27	17.22
Ecological farmland	2.56	0.50	50.24	16.08
Conventional farmland	1.78	0.45	40.02	13.50

rhiza) which converts the unavailable forms of the mineral to forms which are plant-available. Table 9.11 shows these results.

The figure of 50 ppm is equivalent to about 125 kg of the material per hectare. True, with the present methods of non-return of faeces to the soil, there is still a "mining" of the farm. This, however, is less than the "mining" seen in chemical agriculture. The situation in respect of potassium is more serious both for the chemical and ecological farmer. The potassium so much required by the plant is, as has been stated, mainly rejected by all members of the animal kingdom and is eventually lost to the plant/animal/food cycle by finding its way to the sea. The work at Nagele shows that the use of green manures would help reduce the loss of potassium. The technique would reduce the considerable leaching loss from bare ground. Nevertheless one has to face up to the fact that in ecological and chemical farming systems, where crop sales are a part of the farm income, there is a distinct possibility that the farm is in negative potassium balance. Ecologically the balance is corrected by the use of rock potassic salts, or where available cement kiln dust. The work at Haughley has shown that these materials, under ecological methods, release potassium to the soil. The foregoing concerning the losses of both potassium and phosphorus must exercise men's minds. The only real solution is to follow the Chinese method of ensuring that all waste products are recycled. Methods will need to be found to stop the contamination of these waste products with heavy industrial metals, which are rejected by all farming authorities.

9.6 Pollution and Animal Welfare

Modern opinion is turning against the large intensive livestock unit, and already in Europe there has been rejection of overcrowding in hen battery cages. Intensive pig and cattle units are being opposed, not only because of animal overcrowding and questionable husbandry practices, but also because of smell pollution. Soon no farming system will be able to continue when the urban voter uses his superior voting numbers to stop what he or she considers cruel or polluting. Pollution problems are not restricted to smell and noise alone – there is the problem of pollution caused by the leaching of plant nutrients to watercourses. In particular there is concern at the high nitrite levels in water, which are not helped by the use of nitrogenous fertilizers. The nitrates in the waterways give more than enough nutrient to allow excessive growth of water weeds, which then cause blockage of rivers. Clearly, therefore, holistic farming, which pays great attention to animal welfare and pollution, is at a great advantage in present-day political thought. It must, however, be emphasized that the uncaring organic farmer who uses manures direct from hen-battery cages without ensuring that the free nitrogenous salts have been composted with straw is just as great a source of pollution as his chemical counterpart.

9.7 Labour Usage

It has been shown that ecological farming requires about 16% more labour than conventional methods.[14] This in itself can cause problems not only in finding suitable staff, but also in the financing of the extra staff. At Haughley the extra labour charge amounts to about £23 per hectare per year, but because the gross margins are so much higher in the ecological systems this charge does not reduce the profit to below that of chemical farms. However, the development of the microchip and the advent of factory robots does mean that the high-technology countries of the west are faced with the distinct possibility that high long-term unemployment is going to be a built-in aspect of society. Baxter in his studies[15] showed that part of the energy deficiency in U.K. agriculture could be traced to the replacement of the blue-collar worker on farms by larger, more powerful, and more fuel-consuming tractors and other mechanised equipment. A British farm worker today costs in total a minimum of about £120 per week. Governments are notorious at not publishing the true costs of an unemployed person, but in 1984 a figure of about £95 per week was considered correct. Perhaps government policy should be directed at methods of returning labour to primary food production, which if it did nothing else would certainly reduce the energy gap, make agriculture more sustainable, and give useful employment to many people. This consideration is the basis of real holistic thinking.

References

1. *Quarterly Review of Soil Association*, March 1977.
2. Lockeretz, W. *et al.* (1975) *Organic and Conventional Crop Production in the Corn Belt*. Center for the Biology of Natural Systems, Washington University.
3. Kowalski, R. (1982) Personal communication.
4. Lockeretz, W. *et al.* (1984) Comparison of conventional and organic farming in the corn belt. In: Bezdicek, D. F. *et al.* (eds), *Organic Farming: Current Technology and its Role in a Sustainable Agriculture*. American Society of Agronomy, Special Publication No. 46, Madison, Wisconsin, U.S.A.
5. Oelhaf, R. C. (1978) *Organic Agriculture*. John Wiley, Chichester.
6. Vine, and Batemen, (1981) Organic farming systems in England and Wales: practice, performance and implications.
7. Bateman, and Lampkin, (1986) Economics of organic farming. Paper presented at the University College of Wales, Agricultural Society, November 1986. (Originally published in *Agricultural Administration* – forthcoming).

8. Schuphan, W. (1974) Nutritional values of crops as influenced by organic and inorganic fertilisers. *Qual. Plant*, **23**, 333–58.
9. Vogtmann, H. (1981) *The quality of agricultural produce originating from different systems of cultivation.* Translated from German (published 1979) by Soil Association.
10. Nelson, M. *et al.* (1985) Family food purchase and home food consumption; a comparison of nutrient contents. *Br. J. Nutr.*, **54**, 373–87.
11. Blaxter, K. L. (1975) The energetics of British agriculture. *J. Sci. Fd Agric.*, **26**, 1055–64.
12. Personal communication in 1982 to the author by Director of Research at the Dutch Experimental Station, Nagele.
13. Anon (1977) Organic materials and soils productivity. F.A.O. Rome, *Bulletin 35*.
14. Anon (1975) *Changes in farm production and efficiency: a summary*.
15. Blaxter, K. L. (1974) Power and the agricultural revolution. *New Sci.*, **61**(885), 400–3.

Epilogue

HOLISTIC agriculture is shown to be capable of replacing the more conventional methods used by chemical farmers. What is certain is that this new approach to farming is not just a playchild of the wealthy, nor is it a return to the farming employed by our ancestors; it is a system which takes all the modern knowledge of the agricultural and other natural scientists, and develops practical methods of putting the knowledge into practice, ideally with no damage to the environment, but at the worst with the very minimum of harm. However, it is not without problems, which need research by the learned establishments.

Finally in the last analysis in a democratic society a farmer is free to farm as he wishes. He should not, however, feel free to harm himself or others. In my opinion harm means: harm to the unborn by his profligate use of unrenewable resources, harm to his neighbour by causing pollution or environmental damage; and perhaps most important of all harm to himself by subjecting all living things to stresses to satisfy his greed, when this can be avoided by his adopting a holistic approach to his farming.

Appendix I

TABLE 1 *Approximate percentage dry matter, percentage nitrogen, and carbon:nitrogen ratios of some compostable materials*

Material	Dry matter (%)	Nitrogen (%)	C:N ratio
Urine	5.00	16.00	N
Fish scraps	50.00	8.50	N
Poultry manure	50.00	6.30	N
Meat scraps	40.00	5.10	N
Sheep/goat manure	35.00	3.75	N
Pig manure	20.00	5.10	N
Horse manure	25.00	2.30	N
Cow manure	20.00	1.70	N
Slaughterhouse waste	50.00	8.50	2:1
Blood	15.00	12.00	3:1
Activated sludge	30.00	5.50	6:1
Night soil	30.00	6.00	8:1
Young grass clippings	15.00	4.00	12:1
Cabbage leaves	15.00	3.60	12:1
Tomatoes	20.00	3.30	12:1
Farmyard manure	20.00	2.20	14:1
Onions	20.00	2.65	16:1
Ensiled grass waste	20.00	2.70	18:1
Lucerne	22.00	2.70	19:1
Old grass clippings	20.00	2.30	19:1
Turnip tops	12.00	2.30	19:1
Ragwort	20.00	2.20	21:1
Potato haulms	23.00	1.50	25:1
Mixed garbage	20.00	2.20	25:1
Mustard (plants)	15.00	1.50	26:1
Red clover	20.00	1.80	27:1
Carrots	20.00	1.60	27:1
Seaweed	15.00	1.90	29:1
Fern (bracken)	25.00	1.20	43:1
Whole turnips	9.00	1.00	44:1
Oat straw	85.00	1.10	48:1
Flax waste/linen	15.00	0.90	58:1
Wheat straw	85.00	0.30	100:1
Sawdust	80.00	0.10	500:1
Newspaper	95.00	0.00	C
Kraft (brown) paper	95.00	0.00	C

N These materials can be considered for composting purposes to contain no organic carbon.
C These materials can be considered for composting purposes to contain no nitrogen.

TABLE 2   *The chemical composition of some common foodstuffs*

Foodstuff	Oil (%)	Protein (%)	Fibre (%)	Calcium (%)	Phosphorus (%)	Sodium (%)	TDN Pigs	ME Poultry (MJ/kg)	ME Rumen (MJ/kg)
Cereals									
Barley	1.79	10.90	5.00	0.05	0.37	0.04	72.80	11.76	13.70
Maize	3.50	8.70	3.10	0.05	0.35	0.02	77.10	14.29	14.20
Maize, flaked	2.00	9.00	1.50	0.01	0.30	0.02	83.00	15.50	15.00
Oats	5.00	10.50	10.50	0.10	0.35	0.04	63.00	10.60	11.50
Rice	2.00	8.00	8.00	0.08	0.32	0.04	75.50	11.10	13.50
Rye	1.70	12.00	2.20	0.06	0.47	0.02	75.00	11.80	13.75
Sorghum/milo	3.20	11.00	2.10	0.02	0.31	0.02	76.00	13.60	13.30
Wheat	1.42	11.30	2.50	0.04	0.38	0.04	78.40	13.40	14.02
Vegetable proteins									
Beans	1.10	26.52	7.20	0.13	0.40	0.01	67.00	11.50	12.60
Groundnut cake (exp.)	5.90	51.90	7.20	0.25	0.55	0.04	80.00	15.30	12.73
Linseed cake (exp.)	7.40	33.42	9.70	0.40	0.80	0.06	75.00	7.19	12.86
Lupins	10.00	36.00	10.93	0.20	0.30	0.03	90.00	13.02	13.42
Peas	1.40	19.30	5.21	0.13	0.32	0.01	70.00	10.70	12.00
Soya meal (ext.)	2.00	43.00	9.10	0.29	0.64	0.02	69.00	9.22	11.00
Soya meal (dehulled)	1.77	48.10	3.50	0.23	0.97	0.04	73.00	10.20	13.22
Soya beans	16.50	36.50	5.30	0.20	0.60	0.02	91.00	13.80	15.20
Sunflower ext. meal	1.40	30.00	23.00	0.90	0.70	0.04	59.00	7.84	12.70
Animal proteins									
Blood meal	1.00	90.00	0.50	0.30	0.25	0.55	63.00	13.10	13.90
Fish meal (white fish)	4.10	66.00	1.00	4.20	3.50	1.61	66.00	12.45	11.10
Herring meal	6.83	71.00	0.00	3.00	2.20	2.20	74.00	12.90	12.33
Meat meal	4.90	60.00	0.00	5.50	3.00	0.50	60.00	10.60	13.36
Meat and bone meal	9.86	45.00	1.50	10.50	5.10	0.59	55.00	8.30	10.51

Foodstuff	Oil (%)	Protein (%)	Fibre (%)	Calcium (%)	Phosphorus (%)	Sodium (%)	TDN Pigs	ME Poultry (MJ/kg)	ME Rumen (MJ/kg)
Milk products									
Dried buttermilk	7.00	35.30	0.00	1.40	0.98	0.52	96.00	11.34	14.40
Dried skim milk	1.50	34.00	0.00	1.28	1.04	0.50	88.00	11.53	13.70
Dried whey	1.50	12.50	0.00	0.90	0.70	0.90	86.00	11.34	14.00
Cereal by-products									
Maize germ meal	7.00	11.50	4.00	0.08	0.33	0.03	82.00	12.91	14.89
Maize gluten meal	3.00	23.50	4.00	0.10	0.33	0.04	72.00	9.20	13.40
Oats groats	6.00	13.00	1.50	0.04	0.40	0.04	86.00	12.30	13.40
Oat feed	2.40	5.50	28.00	0.20	0.12	0.05	40.00	4.61	8.50
Rice bran (ext.)	0.80	14.20	15.60	0.17	1.41	0.02	61.00	7.50	10.80
Wheat bran	4.00	14.70	11.00	0.14	1.22	0.06	58.00	8.30	9.90
Wheatfeed	4.00	15.00	6.00	0.09	0.66	0.05	65.00	10.10	11.00
Brewer's/distiller's waste									
Brewer's grains (dried)	6.50	18.50	15.00	0.28	0.15	0.06	60.00	7.65	11.80
Distillery dried sols.	9.30	27.00	7.20	1.00	0.80	0.10	81.00	12.73	14.10
Distiller's maize grains	6.50	26.30	9.30	0.25	0.90	0.05	73.00	11.99	13.08
Malt culms	1.30	21.20	14.30	0.05	0.37	0.04	65.00	11.79	11.23
Yeast (dried)	0.70	44.00	2.50	0.10	1.60	0.02	73.00	7.84	11.62
Fats									
Corn/soya oil	98.00	—	—	—	—	—	211.00	37.00	35.20
Tallow	98.00	—	—	—	—	—	222.00	39.28	36.70

Note: 1 kg of dried butter milk or dried skim milk is equivalent to 10 litres of the liquid product.
1 kg of dried whey is equivalent to 15.25 litres of liquid whey.

(continued)

TABLE 2 continued

Foodstuff	Oil (%)	Protein (%)	Fibre (%)	Calcium (%)	Phosphorus (%)	Sodium (%)	TDN Pigs	ME Poultry (MJ/kg)	ME Rumen (MJ/kg)
Miscellaneous									
Apple pommace	2.00	6.00	21.00	0.05	0.17	0.00	50.00	2.67	7.05
Sweet biscuit meal	17.90	6.30	0.35	0.09	0.20	0.31	94.00	17.01	14.47
Low oil biscuit meal	4.13	11.70	1.00	0.13	0.13	0.65	84.00	14.80	12.50
Bread	0.50	1.80	2.50	0.13	0.10	0.03	82.00	14.73	–
Cassava/manioc	0.50	1.80	2.50	0.13	0.10	0.03	80.00	1.50	15.30
Citrus pulp	3.50	6.00	12.00	2.00	0.01	0.09	45.00	5.50	6.00
Grape residue	4.00	14.50	30.00	0.35	0.20	0.15	30.00	4.00	3.13
Grass meal	3.00	15.00	20.00	1.00	0.22	0.31	40.00	5.53	9.24
Lucerne meal	2.50	20.00	18.00	1.70	0.30	0.30	45.00	6.90	9.30
Sugar beet pulp dried	0.60	8.90	18.30	0.85	0.08	0.03	65.00	2.60	11.50
Swill (meal equivalent)	11.90	15.64	3.40	0.34	0.68	0.27	85.00	13.60	12.20
Molasses	0.00	3.50	0.00	0.11	0.02	0.83	60.00	8.30	12.90
Potato crisps	33.00	8.00	0.50	0.10	0.10	2.20	136.00	22.00	17.50
Seaweed, dried	4.00	5.00	9.00	1.10	0.05	1.00	33.00	5.50	7.50
Minerals									
Steamed bone flour	4.00	6.00	0.00	32.00	13.00	–	–	–	–
Limestone flour – oyster	–	–	–	39.50	–	–	–	–	–
Salt	–	–	–	–	–	39.30	–	–	–

The data to compile Tables 2 and 3 were provided by Vitrition Ltd. (animal food supplement manufacturers), Stamford, Lincolnshire. I express my gratitude to the company for providing this information.

TABLE 3 *Major amino acids, linoleic acid, and undegradable (by-pass) protein (UDP) content of some common foodstuffs*

Foodstuff	Total lysine (%)	Available lysine (%)	Methionine (%)	Methionine + cystine (%)	Linoleic acid (%)	UDP (%)
Cereals						
Barley	0.34	0.32	0.17	0.33	0.90	1.63
Maize	0.22	0.18	0.17	0.33	2.10	4.30
Maize, flaked	0.20	0.17	0.16	0.31	0.80	4.50
Oats	0.40	0.38	0.16	0.41	2.30	2.10
Rice	0.28	0.23	0.17	0.47	0.70	1.60
Rye	0.46	0.43	0.14	0.42	1.05	2.40
Sorghum/milo	0.28	0.23	0.16	0.31	1.24	5.50
Wheat	0.38	0.35	0.20	0.42	1.00	1.78
Vegetable proteins						
Beans	1.90	1.40	0.26	0.54	0.53	5.30
Groundnut cake (exp.)	1.40	1.00	0.46	1.06	0.90	10.38
Linseed cake (exp.)	1.10	0.08	0.82	1.45	3.01	13.40
Lupins	4.50	3.80	0.19	1.10	2.35	19.80
Peas	1.90	1.30	0.20	0.50	0.45	3.86
Soya meal (ext.)	2.80	2.32	0.73	1.32	0.69	17.22
Soya meal (dehulled)	3.10	2.70	0.88	1.47	0.75	19.25
Soya beans	2.40	1.90	0.61	1.11	9.10	14.60
Sunflower ext. meal	1.00	0.90	1.27	1.83	0.80	8.20
Animal proteins						
Blood meal	9.50	6.00	1.00	2.50	0.00	41.60
Fish meal (white fish)	6.30	4.20	2.00	2.80	0.40	39.60
Herring meal	7.00	5.30	2.10	2.70	0.70	49.70
Meat meal	4.14	3.60	1.32	1.86	0.14	30.00
Meat and bone meal	2.63	1.90	0.78	1.27	0.20	22.50

(*continued*)

TABLE 3 *continued*

Foodstuff	Total lysine (%)	Available lysine (%)	Methionine (%)	Methionine + cystine (%)	Linoleic acid (%)	UDP (%)
Milk products						
Dried buttermilk	2.70	2.40	0.88	1.21	0.00	3.53
Dried skim milk	2.60	2.30	0.87	1.30	0.00	3.40
Dried whey	0.90	0.75	0.27	0.58	0.00	1.25
Cereal by-products						
Maize germ meal	0.30	0.23	0.24	0.47	4.92	2.30
Maize gluten meal	0.30	0.23	0.24	0.47	1.65	4.70
Oat groats	0.61	0.50	0.22	0.40	2.76	5.00
Oat feed	0.05	0.04	0.00	0.00	1.10	1.24
Rice bran (ext.)	0.60	0.10	0.12	0.32	0.50	3.50
Wheat bran	0.59	0.38	0.17	0.46	0.20	4.41
Wheatfeed	0.60	0.40	0.69	1.59	0.20	4.65
Brewer's/distiller's waste						
Brewer's grains (dried)	0.49	0.38	0.31	0.61	3.50	5.55
Distillery dried sols.	0.80	0.62	0.50	0.82	2.60	8.10
Distiller's maize grains	0.90	0.85	0.60	0.92	0.00	7.89
Malt culms	0.83	0.67	0.38	0.74	1.06	8.15
Yeast (dried)	2.81	2.41	0.69	1.59	0.20	8.80
Fats						
Corn/soya oil	—	—	—	—	50.00	—
Tallow	—	—	—	—	3.40	—

Foodstuff	Total lysine (%)	Available lysine (%)	Methionine (%)	Methionine + cystine (%)	Linoleic acid (%)	UDP (%)
Miscellaneous						
Apple pommace	0.40	0.31	0.13	0.18	0.00	1.60
Sweet biscuit meal	0.35	0.22	0.20	0.39	1.15	2.52
Low oil biscuit meal	0.21	0.11	0.13	0.27	0.00	4.68
Bread	0.60	0.40	0.15	0.32	0.16	4.68
Cassava/manioc	0.06	0.02	0.01	0.02	0.00	0.36
Citrus pulp	0.20	0.16	0.10	0.20	0.00	1.20
Grape residue	–	–	–	–	–	3.00
Grass meal	0.70	0.41	0.50	0.55	0.75	4.50
Lucerne meal	0.94	0.68	0.17	0.39	0.00	6.00
Sugar beet pulp, dried	0.60	0.40	0.01	0.01	0.00	2.67
Swill (meal equivalent)	0.60	0.40	0.00	0.00	0.00	3.20
Molasses	0.10	0.05	0.05	0.09	0.00	1.05
Potato crisps	0.30	0.20	0.00	0.00	4.20	2.40
Seaweed dried	0.29	0.20	0.06	0.12	0.00	1.00

TABLE 4　*The composition (on dry matter) of some common fodder crops*

Foodstuff	Dry matter (%)	Protein (%)	UDP (%)	Calcium (%)	Phosphorus (%)	ME Rumen (MJ/kg)	TDN Pigs (MJ/kg)
Straws							
Oat straw	86.00	3.80	0.75	0.09	0.03	7.00	41.00
Barley straw	86.00	3.00	0.96	0.22	0.06	6.60	28.00
Root break crops							
Cabbage	11.00	13.60	2.72	0.14	0.07	10.40	68.00
Fodder beet	23.00	11.80	2.76	0.24	0.09	13.70	82.00
Kale	14.00	15.70	3.14	0.28	0.06	11.10	65.00
Mangolds	12.50	9.10	1.82	0.01	0.04	12.50	79.00
Potatoes	24.00	9.00	1.80	0.02	0.08	12.50	83.00
Fodder rape	14.00	20.00	4.00	0.28	0.08	9.50	64.00
Swede/turnip	9.00	11.40	2.31	0.06	0.04	11.70	73.00
Silage and hay							
Meadow hay	85.00	8.50	2.10	0.40	0.30	8.40	—
Lucerne hay	85.00	19.30	4.83	0.48	0.30	8.30	—
Grass silage	20.00	16.00	4.00	0.48	0.27	9.90	—
Legume silage	22.00	20.50	5.10	0.45	0.27	9.95	—

Appendix II

TABLE 1 *Suggested nutrient standards for hens*

Nutrient	Chick		Grower		Layer	
	Min.	Max.	Min.	Max.	Min.	Max.
Protein (%)	17.50	18.50	14.00	15.00	16.00	17.00
Total lysine (%)	0.90	1.00	0.60	–	0.80	–
Methionine (%)	0.35	–	0.25	–	0.33	–
Methionine + cystine (%)	0.67	–	0.50	–	0.66	–
Calcium (%)	0.60	0.80	0.50	1.10	3.20	4.00
Total phosphorus (%)	0.50	0.70	0.50	0.65	0.50	0.55
Sodium (%)	0.12	0.19	0.12	0.19	0.12	0.19
Metabolizable energy (MJ/kg)	11.60	11.91	11.30	11.53	11.62	11.72
Linoleic acid (%)	0.75	–	0.70	–	1.20	1.35
Fibre (%)	2.00	3.50	–	5.50	2.50	–
Raw material constraints						
Animal protein (%)	2.50	–	1.25	–	1.25	–
Maize gluten meal (%)	–	0.00	–	2.50	–	5.00
Wheatfeed/oatfeed (%)	–	12.50	–	7.50	–	15.00

Vitamin supplementation is required for stock housed indoors.

TABLE 2 Suggested nutrient standards for ducks and geese

Nutrient	Starter		Grower		Finisher	
	Min.	Max.	Min.	Max.	Min.	Max.
Protein (%)	20.00	21.00	17.00	19.00	14.00	15.50
Total lysine (%)	1.00	1.10	0.75	0.85	0.60	0.70
Methionine (%)	0.40	—	0.30	—	0.28	—
Methionine + cystine (%)	0.75	—	0.60	—	0.55	—
Calcium (%)	1.00	1.10	0.90	1.00	0.80	1.00
Total phosphorus (%)	0.65	—	0.65	—	0.65	—
Sodium (%)	0.12	0.16	0.12	0.16	0.12	0.16
Metabolizable energy (MJ/kg)	11.80	12.60	11.50	12.00	12.30	12.50
Linoleic acid (%)	0.75	—	0.75	—	0.70	—
Raw material constraints						
Animal protein (%)	2.50	—	1.25	—	1.25	—

Vitamin supplementation is required for stock housed indoors.

TABLE 3   *Suggested nutrient standards for turkeys*

Nutrient	Starter		Grower		Finisher	
	Min.	Max.	Min.	Max.	Min.	Max.
Protein (%)	26.00	28.00	17.00	20.00	16.00	18.00
Total lysine (%)	1.55	1.65	1.20	1.25	0.80	0.90
Methionine (%)	0.50	0.60	0.38	0.40	0.33	0.36
Methionine + cystine (%)	0.90	1.00	0.76	0.80	0.66	0.70
Calcium (%)	1.00	1.20	1.00	1.20	0.90	1.00
Total phosphorus (%)	0.85	0.95	0.80	0.90	0.75	0.90
Sodium (%)	0.15	0.19	0.15	0.19	0.15	0.19
Metabolizable energy (MJ/kg)	11.75	12.15	12.00	12.50	12.00	12.50
Linoleic acid (%)	1.00	—	0.90	—	0.80	—
Raw material constraints						
Animal protein (%)	5.00	—	2.50	—	2.50	—

Vitamin supplementation is required.
Only use good-quality materials in turkey diets.

TABLE 4 *Suggested nutrient standards for pigs*

Nutrient	Sows		Creep		Grower		Finisher	
	Min.	Max.	Min.	Max.	Min.	Max.	Min.	Max.
Oil (%)	2.50	—	—	—	—	—	—	—
Protein (%)	15.00	15.50	20.00	—	18.00	—	16.50	17.50
Available lysine (%)	0.70	—	1.20	—	1.10	—	0.70	—
Calcium (%)	0.80	0.95	0.90	1.00	0.68	0.80	0.70	0.90
Total phosphorus (%)	0.60	0.90	0.65	0.75	0.50	0.75	0.50	0.80
Sodium (%)	0.16	0.20	0.16	0.20	—	0.35	—	0.25
TDN	70.00	73.00	75.00	77.00	77.00	79.00	75.00	76.00
Fibre (%)	4.00	5.00	3.00	4.50	4.50	5.00	2.50	—
Raw material constraints								
Animal protein (%)	1.25	—	5.00	—	2.50	—	—	—
Wheatfeed/oatfeed (%)	7.50	—	5.00	10.00	5.00	15.00	—	20.00

Dry matter content of pig diets should not be less than 25%.

TABLE 5   *Suggested daily nutrient standards for dairy cows*

(1) Friesians

Milk yield (kg)	Gross energy metabolizable (q) = 0.6					Gross energy metabolizable (q) = 0.7					
	5	10	15	20	30	5	10	15	20	30	40
Rumen-degraded protein (g)	620	815	1010	1210	1615	590	770	960	1145	1535	1930
Undegraded (by-pass) protein (g)	—	65	190	320	565	—	95	235	370	630	885
Calcium (g)	36	49	63	77	105	36	49	63	77	105	133
Total phosphorus (g)	25	28	36	44	60	25	28	36	44	60	75
Metabolizable energy (MJ)	79	104	129	155	207	76	99	123	147	197	248

Dry matter intake   18 kg   18 kg

Dry matter intake as roughage food   7.2 kg   7.2 kg

(2) Ayrshires

Milk yield (kg)	Gross energy metabolizable (q) = 0.6					Gross energy metabolizable (q) = 0.7					
	5	10	15	20	30	5	10	15	20	30	40
Rumen-degraded protein (g)	575	770	975	1180	1605	545	735	925	1120	1520	1935
Undegraded (by-pass) protein (g)	—	110	245	380	640	—	140	285	430	705	970
Calcium (g)	33	46	60	74	102	33	46	60	74	102	130
Total phosphorus (g)	22	27	35	43	59	22	27	35	43	59	75
Metabolizable energy (MJ)	74	99	125	151	206	70	94	119	144	195	248

Dry matter intake   16 kg   16 kg

Dry matter intake as roughage food   6.4 kg   6.4 kg

(3) Jerseys

Milk yield (kg)	Gross energy metabolizable (q) = 0.6				Gross energy metabolizable (q) = 0.7				
	5	10	15	20	5	10	15	20	30
Rumen-degraded protein (g)	545	775	1010	1245	520	735	955	1185	1650
Undegraded (by-pass) protein (g)	25	180	335	485	45	215	375	535	840
Calcium (g)	35	46	62	78	35	46	62	78	110
Total phosphorus (g)	20	29	40	51	20	29	40	51	72
Metabolizable energy (MJ)	70	99	129	160	67	94	123	152	212
Dry matter intake	15 kg				15 kg				
Dry matter intake as roughage food	6 kg				6 kg				

After CAB Nutrient Requirements of Livestock (1984), by kind permission of Agricultural and Food Research Council.

TABLE 6   *Suggested daily nutrient requirements for growing cattle*

Gross energy metabolizable (q) = 0.5

	Daily liveweight gain (kg)		
	0.25	0.50	0.75
100 kg liveweight			
Appetite (kg of dry matter)	4.00	4.00	4.00
Roughage (kg of dry matter)	2.00	2.00	2.00
Rumen-degraded protein (g)	160.00	200.00	250.00
Undegraded (by-pass) protein (g)	15.00	60.00	90.00
Calcium (g)	7.00	12.00	16.50
Total phosphorus (g)	4.00	6.25	8.50
Metabolizable energy (MJ)	21.00	26.00	32.00
200 kg liveweight			
Appetite (kg of dry matter)	6.00	6.00	6.00
Roughage (kg of dry matter)	4.50	4.50	4.50
Rumen-degraded protein (g)	255.00	310.00	375.00
Undegraded (by-pass) protein (g)	—	—	—
Calcium (g)	9.00	14.00	19.00
Total phosphorus (g)	5.50	7.75	10.25
Metabolizable energy (MJ)	33.00	39.00	48.00
300 kg liveweight			
Appetite (kg of dry matter)	8.00	8.00	8.00
Roughage (kg of dry matter)	6.00	6.00	6.00
Rumen-degraded protein (g)	335.00	400.00	485.00
Undegraded (by-pass) protein (g)	—	—	—
Calcium (g)	12.00	17.00	22.00
Total phosphorus (g)	7.00	9.25	11.50
Metabolizable energy (MJ)	43.00	52.00	62.00
400 kg liveweight			
Appetite (kg of dry matter)	9.50	9.50	9.50
Roughage (kg of dry matter)	6.25	6.25	6.25
Rumen-degraded protein (g)	405.00	485.00	590.00
Undegraded (by-pass) protein (g)	—	—	—
Calcium (g)	14.00	19.00	24.00
Total phosphorus (g)	11.00	15.00	18.00
Metabolizable energy (MJ)	52.00	62.00	75.00
500 kg liveweight			
Appetite (kg of dry matter)	11.00	11.00	11.00
Roughage (kg of dry matter)	7.25	7.25	7.25
Rumen-degraded protein (g)	475.00	565.00	685.00
Undegraded (by-pass) protein (g)	—	—	—
Calcium (g)	17.00	21.00	26.00
Total phosphorus (g)	14.00	17.00	20.00
Metabolizable energy (MJ)	61.00	73.00	88.00

Gross energy metabolizable $(q) = 0.6$

	Daily liveweight gain (kg)		
	0.25	0.50	0.75
100 kg liveweight			
Appetite (kg of dry matter)	4.00	4.00	4.00
Roughage (kg of dry matter)	2.00	2.00	2.00
Rumen-degraded protein (g)	150.00	185.00	225.00
Undegraded (by-pass) protein (g)	25.00	75.00	115.00
Calcium (g)	4.00	6.25	8.50
Total phosphorus (g)	4.00	6.25	8.50
Metabolizable energy (MJ)	20.00	24.00	28.00
200 kg liveweight			
Appetite (kg of dry matter)	6.00	6.00	6.00
Roughage (kg of dry matter)	4.50	4.50	4.50
Rumen-degraded protein (g)	240.00	285.00	335.00
Undegraded (by-pass) protein (g)	—	10.00	30.00
Calcium (g)	9.00	14.00	19.00
Total phosphorus (g)	5.50	7.75	10.25
Metabolizable energy (MJ)	31.00	36.00	43.00
300 kg liveweight			
Appetite (kg of dry matter)	8.00	8.00	8.00
Roughage (kg of dry matter)	6.00	6.00	6.00
Rumen-degraded protein (g)	315.00	370.00	435.00
Undegraded (by-pass) protein (g)	—	—	—
Calcium (g)	12.00	17.00	22.00
Total phosphorus (g)	7.00	9.25	11.50
Metabolizable energy (MJ)	40.00	48.00	56.00
400 kg liveweight			
Appetite (kg of dry matter)	9.50	9.50	9.50
Roughage (kg of dry matter)	6.25	6.25	6.25
Rumen-degraded protein (g)	380.00	450.00	530.00
Undegraded (by-pass) protein (g)	—	—	—
Calcium (g)	14.00	19.00	24.00
Total phosphorus (g)	11.00	15.00	18.00
Metabolizable energy (MJ)	49.00	58.00	68.00
500 kg liveweight			
Appetite (kg of dry matter)	11.00	11.00	11.00
Roughage (kg of dry matter)	7.25	7.25	7.25
Rumen-degraded protein (g)	445.00	525.00	615.00
Undegraded (by-pass) protein (g)	—	—	—
Calcium (g)	17.00	21.00	26.00
Total phosphorus (g)	14.00	17.00	20.00
Metabolizable energy (MJ)	57.00	67.00	79.00

After *CAB Nutrient requirements of Livestock – Ruminants* (1984), by kind permission of Agricultural and Food Research Council.

TABLE 7 *Suggested minimum daily nutrient standards for ewes*

Gross energy metabolizable (*q*) = 0.5

	40 kg ewe with twins			70 kg ewe with single			70 kg ewe with twins		
	133 days pregnant	Term	Lactating	133 days pregnant	Term	Lactating	133 days pregnant	Term	Lactating
Rumen degraded protein (g)	90	110	140	115	135	165	140	170	200
Undegraded (by-pass) protein (g)	5	10	55	–	–	40	–	50	60
Calcium (g)	7	7	10	9	9	11	10	11	14
Total phosphorus (g)	2	2	5	3	3	7	3	4	8
Metabolizable energy (MJ)	12	14	18	15	17	21	18	22	25
Dry matter intake (kg)	1.2	1.2	1.6	1.6	1.6	2.8	1.6	1.6	2.8
Dry matter intake as roughage (kg)	0.8	0.8	1.0	1.0	1.0	1.7	1.0	1.0	1.7

After *CAB Nutrient Requirements of Livestock* (1984), by kind permission of Agricultural and Food Research Council.

TABLE 8 *Suggested daily nutrient standards for goats*

Maintenance requirements including pregnancy

	Stabled or tethered goats (low activity)				Herded goats (high activity)			
Weight of goat (kg)	50	60	70	80	50	60	70	80
Dry matter (kg)	0.95	1.09	1.25	1.35	1.66	1.95	2.15	2.50
Crude protein (g)	75.00	85.00	95.00	105.00	125.00	150.00	165.00	185.00
Calcium (g)	3.00	3.00	4.00	4.00	5.00	6.00	6.00	7.00
Total phosphorus (g)	2.25	2.25	2.75	2.75	3.50	4.25	4.25	5.00
Metabolizable energy (MJ)	11.00	12.00	13.00	14.50	17.00	19.00	22.00	25.00

Additional requirements per day per 1 kg milk

Percentage butterfat	3.50	4.00	4.5
Crude protein (g)	68.00	72.00	77.00
Calcium (g)	2.00	3.00	3.00
Total phosphorus (g)	1.50	2.00	2.25
Metabolizable energy (MJ)	5.00	6.00	7.00

Additional requirements per day for mohair

Annual weight of fleece (kg)	2.00	4.00	6.00	8.00
Crude protein (g)	1.50	2.75	4.00	6.00
Calcium (g)	–	–	–	–
Total phosphorus (g)	–	–	–	–
Metabolizable energy (MJ)	1.50	2.75	4.00	6.00

Daily nutrient requirements for growth

Liveweight (kg)	10		20		30		40		50		60		70		80	
Liveweight gain (g)	50	100	50	100	50	100	50	100	150	200	150	200	150	200	150	200
Crude protein (g)	60	75	70	85	72	85	80	90	90	95	97	105	105	115	120	127
Calcium (g)	2.0	2.0	2.0	2.0	3.0	3.0	3.0	4.0	3.0	4.0	4.0	5.0	5.0	5.0	5.0	5.0
Total phosphorus (g)	1.5	1.5	1.5	1.5	2.25	2.25	2.25	3.0	2.25	3.0	3.0	3.5	3.5	3.5	3.5	3.5
Metabolizable energy (MJ)	5.0	6.25	6.75	8.5	8.5	10.0	10.5	11.5	12.0	13.25	13.5	14.5	14.75	16.25	16.25	17.5

Appendix III

Further Reading

In writing this book I have had to resort to many different sources to find scientific knowledge on subjects of specific interest to the holistic farmer. Many of the papers were published many years ago and have been passed over during the advances made by non-sustainable chemical agriculture, during and since World War II.

Some scientific papers have been abstracted and recorded in the Commonwealth Agricultural Bureau annotated bibliographies. These bibliographies are a rich source of further reading. A complete list of all bibliographies which are available can be obtained from the Bureau at Farnham Royal, Slough, England.

Alas many of the books are now out of print, but can still be found, usually kept in that part of the libraries known as "basement stock".

I have already listed important references at the end of each chapter, but in the following pages I indicate whether the reference is a book, annotated bibliography, or scientific paper published originally in learned journals. Occasionally I give a note concerning the particular book.

I trust that the reader who wishes to study further will find these lists of value.

Chapter 1: Introduction

Books

An Agricultural Testament, Sir Arthur Howard. Oxford University Press, 1940.
The Living Soil, Lady Eve Balfour. Faber & Faber, 1943.
Farmers of Forty Centuries, F. H. King. Republished Rodale, 1980.
(These three classics will stimulate the thoughts of anyone interested in holistic agriculture.)
Small is Beautiful, E. E. Schumacher. Abacus, 1974.
A Guide for the Perplexed, E. F. Schumacher. Abacus, 1978.
(These two books by Schumacher will give the reader a very sound knowledge of the general holistic ideal.)
Energy for World Agriculture, Organic Agriculture, R. C. Oelhaf. John Wiley, 1978.

Annotated Bibliographies

CAB Bibliography NO. SA 1806, *Books on Soil Science, Fertiliser use and plant nutrition*. 324 references.

Chapter 2: Soils and Cultivations

Books

Water for Every Farm, P. A. Yeomans. Murray, 1965.
The Keyline Plan, P. A. Yeomans. Published by the author in 1954. *Water for Every Farm* and *The Keyline Plan* reprinted as *Water for Every Farm* by Murray, 1978; republished by Second Back Row Press, 50 Govett St, Katoomba, N.S.W., Australia. It was P. A. Yeomans who developed the Keyline method of cultivation.

Annotated Bibliography

CAB Bibliography No. S156R (139 references).

Scientific Papers

Low, A. J. (1955) Improvement of the structural state of soils under leys. *J. Soil Sci.*, **6**, 179–97.
Burg, S. P. (1968) Ethylene, plant senescence and abscission. *Pl. Physiol.*, **43**, 1503–11.
Smith, A. M. and Cook, P. J. (1974) Implication of ethylene production by bacteria. *Nature*, **252**, 703–5.
Smith, K. A. and Dowdeswell, R. J. (1974) Field studies of soil atmosphere, relationship between ethylene, oxygen, soil moisture, and temperature. *J. Soil Sci.*, **25**, 217–30.
Lynch, J. M. (1978) Production and phytotoxicity of acetic acid in anaerobic soils containing plant residues. *Soil Biol. Biochem.*, **10**, 131–5.
Smith, A. M. *et al.* (1978) Soil ethylene production triggered by ferrous iron. *Microbial Ecology* (Journal published by Springer Verlag, New York).

Chapter 3: Sources of plant nutrients

Books

Book of Composting. Rodale Press, 1979.
F.A.O. *Soils Bulletin No. 35, Organic Materials and Soil Productivity*.
F.A.O. *Soils Bulletin No. 36, Organic Recycling in Asia*.
F.A.O. *Soils Bulletin No. 40, China: Recycling of Organic Wastes*.
F.A.O. *Soils Bulletin No. 43, Organic Recycling in Africa*.

Annotated Bibliographies

CAB Bibliography No. SB1915, *Waste materials of animal origin (except dung and urine) as fertilisers and manures (1950–1976)*. 92 references.
CAB Bibliography No. SB1914, *Nitrogen fertiliser efficiency as affected by organic manure*. 55 references.
CAB Bibliography No. G401, *Legumes v fertilisers as a source of N*. 103 references.
CAB Bibliography No. S112R, *Town refuse, sewage sludge and ashes as fertilisers (1945–1957)*. 148 references.
CAB Bibliography No. S687R, *Town refuse and sewage sludge (1956–1963)*. 106 references.
CAB Bibliography No. S1634R, *Town refuse and sewage as fertilisers (1963–1973)*. 178 references.
CAB Bibliography No. S1634R, *Town refuse and sewage as fertilisers (1972–1977)*. 134 references.

Scientific Papers

Read, D. J. and Striblet, D. P. (1973) The effect of mycorrhizal infection and phosphorus
  nutrition of ericaceous plants. *Nature*, **244**, 81–2.
Smith, A. M. (1976) Anaerobic microsites in the rhizosphere of plants as mechanisms for
  increasing phosphate availability. Review in *Rural Science Symposium Proceedings*, University of
  New England Armidale, Australia.

Chapter 4: Pests and Diseases

Scientific Papers

Rice, E. L. (1964) Inhibition of N fixing and nitrifying bacteria *Ecology*, **45**, 824–37.
Kiber, R. (1967) Phytotoxicity of plant residues, No. I. *Aust. J. Agric. Res.*, **18**, 361–74.
Turkey, H. B. (1969) Implication of allelopathy in agriculture plant science. *Bot. Rev.*, **35**(1).
Kimber, R. (1973) Phytotoxicity from plant residues, No. II. *Pl. Soil*, **38**, 347–61.
Putnam, A. R. and Duke, W. B. (1978) Allelopathy in agro-ecosystems. *Ann. Rev. Phytopathol.*,
  16, 431–51.
Lovett, J. V. and Sager, G. R. (1978) Influence of bacteria in the phyllosphere of *Camelina sativa*
  on germination of *Linum usitatissimum*. *New Phytol.*, **81**, 617–25.
Rice, E. L. (1979) Allelopathy: an update. *Bot. Rev.*, **45**(1). (This review paper contains a
  bibliography of 405 references.)
Lovett, J. V. and Duffield, A. M. (1981) Allelochemicals of *Camelina sativa*. *J. Appl. Ecol.*, **18**,
  283–90.
Lovett, J. V. *et al.* (1982) Allelopathic activity of cultivated sunflowers. *Proceedings of 10th
  International Sunflower Conference*, pp. 198–204.

Chapter 5: Rotations and Crops

Books

Legume inoculants and their use. F.A.O., Rome, 1984.
Tropical legumes: resource of the future. National Academy of Science, U.S.A., 1979.
Organic Chemicals from the Biomass (editor I. S. Goldstein) CRC Press, Boca Raton Florida, 1981.
Poplars and willows in wood production and land use. F.A.O. Forestry Series No. 10; 1984 revised
  edition.

Annotated Bibliographies

CAB Bibliography No. G361, *Lupinus albus for fodder (1931–1972)*. 250 references.
CAB Bibliography No. G361 (2), *Lupins for seed*. 233 references.
CAB Bibliography No. S1382, *Rotations and productivity (1965–1969)*. 166 references.

Scientific Papers

Stott, K. G. (1956) Cultivation of the basket willow. *Q. Jl Forestry*, April 1956.
Agboola, A. and Fayemi, A. (1971) Preliminary trial on the intercropping of maize with different
  tropical legumes. *J. Agric. Sci. Camb.*, **77**, 219–25.
Agboola, A. and Fayemi, A. (1972a) Fixation and excretion of nitrogen by tropical legumes.
  Agron. J., **64**, 409–12.
Agboola, A. and Fayemii, A. (1972b) Effect of soil management on corn yield and soil nutrients.
  Agron. J., **64**, 641–4.
Brown, H. B. (1935) Effect of soybeans on corn yield. *La Agric. Exp. Stn. Bull.* 265.
Crookston, R. K. (1976) Intercropping: a new version of an old idea. *Crops Soil*, **28**, 7–9.

Dalal, R. C. (1974) Effect of intercropping maize with pigeon peas on grain yield and nutrient uptake. *Exp. Agric.*, **10**, 219–24.
Odland, T. E. (1930) Soybeans for silage and hay. *W. Va. Univ. Agric. Exp. Stn. Bull.* 227.

Chapter 6: Grassland Management

Books

The Clifton Park System of Farming, R. H. Elliot. Republished Faber & Faber, 1943.
(The classic book extolling the virtues of the deep-rooted herbal ley. Elliot first published his results in 1898. The Faber & Faber edition carries a brilliant introduction by the late Sir George Stapldon, founding father of the breeding of grasses for specific purposes.)
Guide to grasses, legumes, herbs, and weeds. Anon. Hunters Seeds Ltd, Chester.

Annotated Bibliographies

CAB CAB Bibliography No. G522, *Sainfoin.* 515 references.
CAB Bibliography No. G175, *Application of sewage sludge and factory effluent to grassland crops.* 48 references.

Chapter 8: Animal Nutrition

Books

Nutrient Standards for Pigs.
Nutrient Standards for Poultry.
Nutrient Standards for Ruminants.
(These three advanced reference books on animal nutrition were the result of the labours of scientific committees appointed by the British Agricultural Research Council to review all scientific papers on animal feeding, and to report their findings as nutrient standards. All are published by the Commonwealth Agricultural Bureau, Farnham Royal, Slough, England. There are similar standards produced by both the American and French Government authorities.)
McCance, R. A. and Widdowson, E. M. *The Composition of Food.* MRC Books published by H.M.S.O., London.
(The book contains analysis details of most human foods, including a considerable number of proprietary products. The analyses are an invaluable source of information to the holistic farmer who feeds livestock on waste products of the human food industry.)

Scientific Papers

Lupins for Livestock (1980) Seminar proceedings, Rutherglen Research Institute, Victoria, Australia.
Animals as waste converters. Proceedings of symposium held at Wageningen in 1983. (Most of the papers are unacceptable to the holistic farmer but a paper dealing with the ensiling of kitchen waste from private houses is of great interest.)

Index